Thickening and Gelling Agents for Food

To Reo

Thickening and Gelling Agents
for Food

Edited by

A. IMESON
Research and Development Manager
Red Carnation Gums Ltd.
London

BLACKIE ACADEMIC & PROFESSIONAL
An Imprint of Chapman & Hall
London · Glasgow · New York · Tokyo · Melbourne · Madras

Published by
Blackie Academic & Professional, an imprint of Chapman & Hall,
Wester Cleddens Road, Bishopbriggs, Glasgow G64 2NZ

Chapman & Hall, 2–6 Boundary Row, London SE1 8HN, UK

Blackie Academic & Professional, Wester Cleddens Road, Bishopbriggs, Glasgow G64 2NZ, UK

Van Nostrand Reinhold Inc., 115 Fifth Avenue, New York NY10003, USA

Chapman & Hall Japan, Thomson Publishing Japan, Hirakawacho Nemoto Building, 6F, 1-7-11 Hirakawa-cho, Chiyoda-ku, Tokyo 102, Japan

DA Book (Aust.) Pty Ltd., 648 Whitehorse Road, Mitcham 3132, Victoria, Australia

Chapman & Hall India, R. Seshadri, 32 Second Main Road, CIT East, Madras 600 035, India

First edition 1992

© 1992 Chapman & Hall

Typeset in 10/12 pt Times New Roman by Best-set Typesetter Ltd., Hong Kong
Printed in Great Britain at the University Press, Cambridge

ISBN   0 7514 0093 9   0 442 30866 3

A catalogue record for this book is available from the British Library

Library of Congress Cataloging-in-Publication data available

Preface

Thickening and gelling agents are invaluable for providing high quality foods with consistent properties, shelf stability and good consumer appeal and acceptance. Modern lifestyles and consumer demands are expected to increase the requirements for these products.

Traditionally, starch and gelatin have been used to provide the desired textural properties in foods. Large-scale processing technology places greater demands on the thickeners and gelling agents employed. Modified starches and specific qualities of gelatin are required, together with exudate and seed gums, seaweed extracts and, most recently, microbial polysaccharides, to improve product mouthfeel properties, handling, and stability characteristics. These hydrocolloids have been established as valuable food additives as a result of extensive practical experience with different products. Nevertheless, the last few years have produced much additional research data from sophisticated new analytical methods. Information on the fine structure of these complex molecules has given a tremendous insight into the three-dimensional conformation of hydrocolloids and their behaviour in solution. Critical components within the biopolymer have been identified which provide particular thickening, suspending, stabilising, emulsifying and gelling properties.

Contributions for this book have been provided by senior development managers and scientists from the major hydrocolloid suppliers in the US and Europe. The wealth of practical experience within this industry, together with chemical, structural and functional data, has been collated to provide an authoritative and balanced view of the commercially significant thickening and gelling agents in major existing and potential food applications.

This is a highly practical manual directed to all people involved in the many diverse aspects of food production. Formulation and development technologists, technical managers, process engineers, production personnel, ingredient purchasers and marketing managers will be able to identify the most appropriate products for preparing high quality foods with consistent properties.

This concise, modern review of hydrocolloid developments will be invaluable as a teaching resource and reference text for all academic and training courses involved with food preparation, production and research.

A.I.

Contents

6 Pectins   124
C.D. MAY

Contributors

Bernard Dalbe   Rhône Poulenc Recherches, Service Applications des Dispersions, 52 rue de la Haie Coq, 93308 Aubervilliers Cedex, France

John E. Fox   G.C. Hahn & Co., Aegidienstrasse 22, 2400 Lubeck, Germany

William Gibson   Kelco International Ltd, Applications Research and Development Department, Waterfield, Tadworth, Surrey, KT20 5HQ, UK

Alan Imeson   Red Carnation Gums Ltd, Sir John Lyon House, 5 High Timber Street, Upper Thames Street, London, EC4V 3PA, UK

Colin D. May   HP Bulmer Pectin Ltd, Plough Lane, Hereford, HR4 0LE, UK

Edvar Onsøyen   Protan Biopolymer A/S, PO Box 494, N-3002, Drammen, Norway

Jan Poppe   Sanofi Bio-Industrie NV, Meulestedekaai, 81 B-9000 Gent, Belgium

André Rapaille   Cerestar, EuroCentre Food, Havenstraat 84, B-1800 Vilvoorde, Belgium

William R. Thomas   FMC Corporation, Marine Colloids Division, 2000 Market Street, Philadelphia, PA19103, USA

Brice Urlacher   Rhône Poulenc Recherches, Service Applications des Dispersions, 52 rue de la Haie Coq, 93308 Aubervilliers Cedex, France

Joast Vanhemelrijck   Cerestar, Eurocentre Food, Havenstraat 84, B-1800 Vilvoorde, Belgium

Roger van Coillie   Aqualon France b.v., BP12, 27640 Alizay, France

David C. Zecher   Aqualon Company, Hercules Research Center, Hercules Road and Route 48, Wilmington, DE19808, USA

1 Alginates

E. ONSØYEN

1.1 Introduction

Alginate is one of the most significant of all the hydrocolloids used in food applications. In order to understand why this is so, this chapter describes those alginate characteristics which are crucial for food applications. In particular, emphasis is placed on how the nature of brown algae used for alginate extraction determines the alginate chemistry and thereby the functional properties and applicability of alginates.

Commercial applications for alginates in foods are based on the interaction between sodium alginate and cations to generate or modify food rheology, usually by the formation of a gel network in the presence of calcium ions. The divalent calcium cations cross-link the alginate polyanionic molecules enabling a gel network to form at any practical temperature. For energy efficiency this is generally carried out in the cold, since alginates are soluble in cold water. Once formed the alginate gel maintains its shape and rheological characteristics throughout thermal processing, including all types of cooking and retorting. Such an alginate network may be freeze–thaw stable, a property which is also a requirement in many food products. By controlling the release of calcium, the setting time of an alginate gel may be adjusted from a few seconds to many minutes and the rheological characteristics manipulated to fulfil the particular specifications of the food product being manufactured.

1.2 Manufacture

1.2.1 Raw materials

Alginate occurs in the cell walls and intercellular spaces of brown algae. The alginate molecules provide both flexibility and strength to the plants, necessary properties adapted to growth conditions in the sea. A diverse range of alginate applications have been developed during the 50 year history of commercial utilisation. The natural properties and functional behaviour of alginates are mimicked in many of these applications.

Brown algae require clean water, with a temperature between 4 and 18°C. Since they are photosynthetic organisms they are restricted to

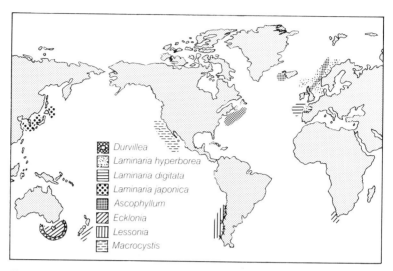

Figure 1.1 Industrially utilised brown algae (from Protan Biopolymer A/S, 1990).

locations with appropriate light conditions, from the tidal zone to a depth of 50 m, depending on the species.

The locations of brown algae industrially utilised for alginate production are shown in Figure 1.1. The brown algae most widely used for industrial production of alginate include *Laminaria hyperborea*, *Laminaria digitata*, *Laminaria japonica*, *Ascophyllum nodosum* and *Macrocystis pyrifera*. In addition, other species of *Laminaria* (*saccharina* and *cloustoni*) as well as some *Ecklonia*, *Lessonia* and *Durvillea* species are used, but on a much smaller scale. However, *Lessonia* and *Durvillea* seem to be of increasing importance as raw materials for alginate production. The alginate application, whether as thickener, stabiliser, gel former or film forming agent generally determines which alginate to choose, and, hence, which raw material, in order to obtain optimal application efficacy. The critical parameters for making the right choice will be discussed in the chemical section.

1.2.2 Commercial alginates

Alginic acid, the free acid form of alginate, is the intermediate product in the commercial manufacture of alginates. Alginic acid has limited stability, like other free acid forms of polysaccharides. In order to make stable water-soluble alginate products, the alginic acid is transformed into the range of commercial alginates by incorporating different salts as shown in Figure 1.2.

Alginic acid

	Na₂CO₃		Na-alginate
	K₂CO₃		K-alginate
	NH₄OH		NH₄-alginate
	Mg(OH)₂		Mg-alginate
	CaCl₂		Ca-alginate
	Propylene oxide		Propylene glycol alginate (PGA)

Figure 1.2 Commercial products from alginic acid.

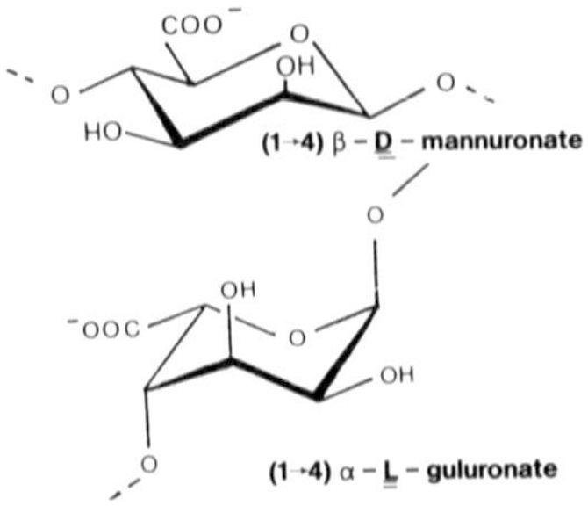

(1→4) β – **D** – mannuronate

(1→4) α – **L** – guluronate

Figure 1.3 The alginate monomer units.

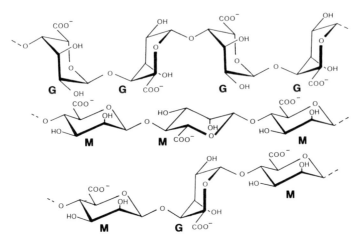

Figure 1.4 Block types in alginate. Top: G-blocks; middle: M-blocks; bottom: MG-blocks.

1.3 Chemical composition

1.3.1 General

Alginates are high molecular weight polymers with regions of flexibility and stiffness. They are salts of alginic acid with a degree of polymerisation usually in the range 100–3000, corresponding to molecular weights of approximately 20 000–600 000. The building blocks of alginic acid are the sugar acid β-D-mannuronic acid (M) and its C-5 epimer, α-L-guluronic acid (G), linked together to form linear molecules with (1,4) glycosidic bonds. Molecular structures are presented in Figures 1.3 and 1.4.

1.3.2 Configuration

It has been shown that the monomeric M- and G-residues in alginates are joined together in sections consisting of homopolymeric M-blocks (MMMMM) and G-blocks (GGGGG) or heteropolymeric blocks of alternating M and G (MGMGMG). In the polymer chain the monomers will tend to find their most energetically favourable structure. For G–G it is the $^1C_4$ chair form linked together with an α-(1,4) glycosidic bond. For M–M it is the $^4C_1$ chair form linked together with a β-(1,4) glycosidic bond. The rather bulky carboxylic group is responsible for an equatorial/equatorial glycosidic bond in M–M, an axial/axial glycosidic bond in G–G and an equatorial/axial glycosidic bond in M–G. The consequence of this is a buckled and stiff polymer in the G-block regions and a flexible ribbon-like polymer in the M-block regions. The MG-block regions have intermediate stiffness.

1.3.3 Alginate biosynthesis

The block structure distribution of an alginate is determined by the biosynthesis of the polymer in seaweed and its subsequent genetic and environmental control. The pathway for alginate biosynthesis in brown algae yields polymannuronate, the homopolymer of M, as an obligate intermediate. An epimerase then acts on the polymer and progresses along the chain effecting the epimerisation from M to G in certain regions of the polymer. As a rule of thumb the transformation from M to G will be more and more complete as the plant tissue grows older. In the stem part of *L. hyperborea* for example, the alginates with the highest gel-forming capacity are found. The stem region possesses tissue with old cells and hence high G content and long G-blocks. In contrast the leaf region possesses tissue which is renewed once a year and therefore has less G content. Early in the growth season of *L. hyperborea* the plants

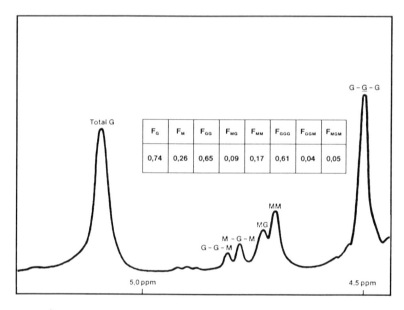

Figure 1.5 ¹H nmr (400 MHz) spectrum of alginate (stipe of *Laminaria hyperborea*) (from Grasdalen, 1983).

possess a double set of leaves before the old leaves from the previous year are rejected.

1.3.4 Block structure analysis

In recent scientific studies, information regarding the detailed chemical composition of alginate has been obtained by nmr (nuclear magnetic resonance) spectroscopy. Figure 1.5 and Table 1.1 show the results for the triplet frequencies and the average block length structures.

1.4 Functional properties

1.4.1 Viscosity

When a water-soluble salt of alginic acid starts to hydrate, the solution gains viscosity. The viscosity is determined by the length of the alginate molecules involved. Commercially available alginates will always contain a range of molecular weight. However, from a practical point of view, they will act according to the Staudinger (Mark Houwink) equation

Table 1.1 Typical composition of different seaweeds (from Protan Biopolymer A/S, 1990).

Algae	M/G	%M	%G	%MM	%MG + GM	%GG	Na-alginate content (%) Raw	Dried
Laminaria hyperborea (stem)	0.45	30	70	18	24	58	3.0–3.5	25–27
Laminaria hyperborea (leaf)	1.20	55	45	36	38	26	2.8–3.8	15–25
Laminaria digitata	1.20	55	45	39	32	29	4	20–26
Ecklonia maxima	1.20	55	45	38	34	28		40
Macrocystis pyrifera	1.50	60	40	40	40	20		26
Lessonia nigrescens	1.50	60	40	43	34	23		35
Ascophyllum nodosum	1.85	65	35	56	18	26	6–8	26–28
Laminaria japonica	1.85	65	35	48	34	18		25
Durvillea antarctica	2.45	71	29	58	26	16		46
Durvillea potatorum	3.35	77	23	69	16	15		53

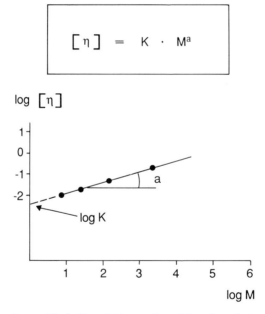

$$[\eta] = K \cdot M^a$$

Figure 1.6 Staudinger (Mark Houwink) equation giving the relation between instrinsic viscosity, [η], and molecular weight, *M*.

giving the relation between intrinsic viscosity and molecular weight (Figure 1.6).

An aqueous solution of alginate has so-called shear thinning or pseudoplastic characteristics. This is a consequence of the length and the stiffness of the hydrated alginate molecules in solution. As long as such a

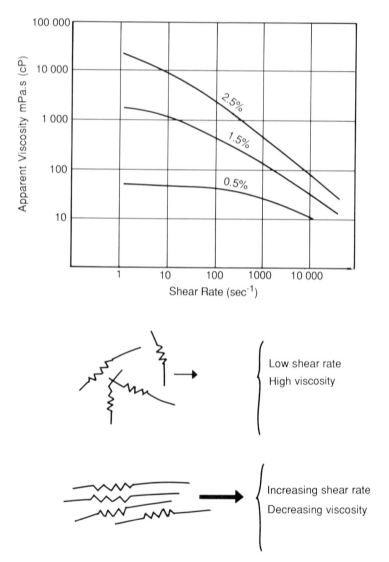

Figure 1.7 Apparent viscosity and shear rate curves for medium viscosity sodium alginate solutions (Cottrell and Kovacs, 1980).

solution is kept under low shear rate conditions, such as during shelf storage or under very slow speed stirring, the different alginate molecules will be directed more or less randomly. By increasing the shear rate however, the molecules will start to direct themselves in a more parallel fashion. This is illustrated in Figure 1.7 and it explains the shear thinning behaviour of an alginate solution.

Table 1.2 Alginate viscosity, mPa s, Brookfield viscometer, model RVT, 20 r.p.m., 20°C.

Concentration (%)	Alginate viscosity grade			
	Low	Medium	High	Very high
0.25	9	15	21	27
0.5	17	41	75	110
0.75	33	93	245	355
1.0	58	230	540	800
1.5	160	810	1950	3550
2.0	375	2100	5200	8750

To meet the target viscosity for a certain application, a choice is made from many different viscosity grades of alginates. Additionally, the alginate concentration can be adjusted to an appropriate level to fulfil special rheological requirements. Table 1.2 shows the viscosity range for various alginates at different concentrations.

In order to increase the viscosity at low alginate concentrations, a small amount of a slightly soluble calcium salt such as calcium sulphate, calcium tartrate or calcium citrate may be added. Calcium ions will react with the alginates to cross-link the molecules and hence increase the molecular weight and viscosity.

1.4.2 Gelation

The main property resulting from the block structure distribution is the ability of alginates to form gels. The main advantage of alginate as a gel former is its ability to form heat-stable gels which can set at room temperatures. In food applications it is primarily the gel formation with calcium ions which is of interest. However, at slightly acid pH, such as for fruit and jam applications, the alginate gel will be an acid type gel or a mixed calcium/acid gel. This is described below in more detail.

In order to react with calcium to form a gel, alginate has to contain a certain proportion of guluronic acid, and the guluronic acid monomers must occur in series. Regions of polyguluronic acid in one alginate molecule are linked to a similar region in another alginate molecule by means of calcium. The gel forming capacity and the resulting gel strength obtained from an alginate is thus very closely connected to the amount of G-blocks and the average G-block length. High G content and long G-blocks give alginates with the highest calcium reactivity and the strongest gel forming potential.

The divalent calcium cation, Ca^{2+}, fits into the guluronic acid structures like an egg in an eggbox, as shown in Figures 1.8 and 1.9 (Grant

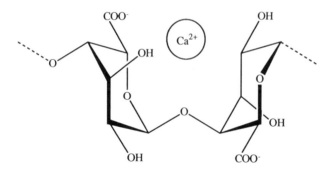

Figure 1.8 Calcium ion position within guluronic acid dimer.

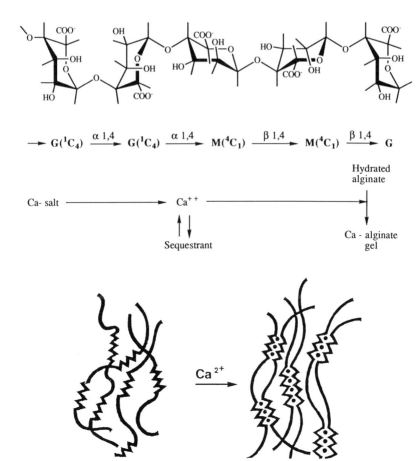

Figure 1.9 Alginate gelation.

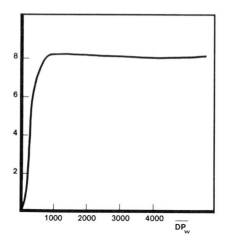

Figure 1.10 Gel strength in a 2% Ca-alginate gel (Fg = 0.38) as a function of weight average degree of polymerisation.

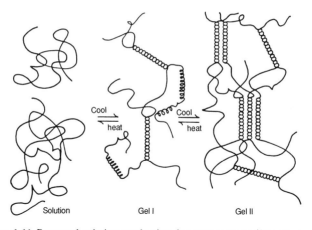

Figure 1.11 Proposed gelation mechanism for carrageenans (from Rees, 1969).

et al., 1973). This may be visualised as a 'molecular cross-linking glue' binding the alginate polymers together by forming junction zones, leading to gelation of the alginate solution. An alginate gel can be considered as part solid and part solution state, the junction zones representing the solid state. After gelation the water molecules are physically entrapped by the alginate matrix or network, but are still free to migrate. This has very important consequences in many applications. The water-holding capacity of the gel is due to capillary forces.

An alginate gel is heat stable. The strength of the gel is broadly

Table 1.3 Alginate gel strength FIRA value (ml).

Alginate origin	Gel strength
L. hyperborea	
Stem	65–75
Whole plant	55–65
Double leaf	45–55
Pure leaf	40–50
A. nodosum	25–35

independent of the molecular chain length, as long as the chains are above a certain length. As shown in Figure 1.10 (Smiderød, 1972), the degree of polymerisation should be above 200 to achieve optimal gel strength.

From a chemical point of view, the formation of a calcium alginate gel should be considered as an ion exchange process. Sodium in sodium alginate, or the cation from any other water-soluble salt of alginate, is exchanged with calcium.

An acid alginate gel will be formed in a similar way by the involvement of junction zones. Here again it is the G-blocks which contribute to the greatest extent to gel formation and gel strength. This can be illustrated by analogy with the well-established model for a two-step gel formation in carrageenans (Figure 1.11). In this model an initial stage of dimerisation is followed by further association to give multiple chain junction zones and greater stability to the three-dimensional gel network.

1.4.3 Gel strength

The alginates extracted from different raw materials may be characterised by their gel strength. As shown in Table 1.3 alginates covering a broad range of gel strength can be produced from a single species of seaweed when different parts of the *L. hyperborea* plant are separated prior to alginate extraction. The FIRA values refer to gel strength measurements using a FIRA jelly tester on alginate gels with very uniform and homogenous consistency (Toft, 1982).

1.5 Gel formation techniques

1.5.1 General

Alginates will form gels with acids and with calcium ions. The gel formation can be controlled through control of the release of calcium or acid into an alginate solution. Both acid and calcium alginate gels are thermoirreversible. A thermoreversible gel can be made under acidic conditions (below pH 4.0 and, preferably, around pH 3.4) by using a combination of alginate and high-methoxy pectin. However, the most important property of alginate as a gel former is its ability to make heat-stable gels in cold systems.

1.5.2 Diffusion setting, neutral pH

In this system, alginate, or an alginate-containing mix, is gelled by being dipped into, or sprayed with, a calcium salt solution. Calcium chloride is used most often. The calcium ions diffuse into the mix containing alginate forming a calcium alginate gel when the calcium ions react with the alginate.

This process is especially suitable for relatively thin or small dimension materials, such as pimiento strips and onion rings, or to provide a thin coating on a product surface. The diffusion process can be increased by raising the concentration of calcium in the setting bath or spray and by using a strongly calcium-reactive alginate, i.e. an alginate with a high proportion of G-blocks.

1.5.3 Diffusion setting, acid pH

In this system, a calcium salt which is insoluble at neutral pH, is mixed with the alginate. When an acid comes into contact with the surface of the mass, the calcium salt is solubilised. The soluble calcium will then react with the alginate and start the gelation process.

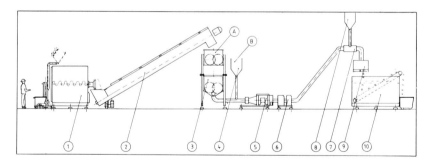

Figure 1.12 Process line for pet food: 1, coarse grinder; 2, screw conveyor; 3, bottom emptying blender with pump; 4, holding tank, alginate/pyrophosphate stock solution; 5, microcutter/emulsifier; 6, in-line transport pump; 7, in-line mixer; 8, holding tank, calcium sulphate stock solution; 9, chunk-forming head; 10, calcium chloride setting bath; A, alginate added dry; B, stock solution of alginate and pyrophosphate (from Protan Biopolymer A/S, 1988).

1.5.4 Internal setting, neutral and acid pH

In this process, calcium is released within the product under controlled conditions. It employs the combination of alginate, a slowly soluble calcium salt and a suitable calcium sequestrant, such as a phosphate or citrate. The sequestrant is needed to bind free calcium and prevent pregelation of the alginate during the time the product is mixed, and before it is cast into its desired shape. The shorter the mixing time, the lower the level of sequestrant needed.

The process may be performed at neutral or acid pH. The acidity may be obtained by the addition of an acidifier, which will accelerate the solubility of calcium salts.

1.5.5 Combined setting

The diffusion and internal setting systems may be combined, thus rapidly providing a gelled outer membrane or coating before the product sets throughout by means of slow calcium release, as in the production of pet food chunks shown in Figure 1.12.

1.5.6 Setting by cooling

In this process, the alginate is dissolved in water together with a low level of a calcium salt and a sequestrant, and kept hot. The high temperature prevents gelation because the alginate chains are in thermal motion, which hinders association of the chains. Setting will start when the solution is cooled and gives a heat-stable calcium alginate gel. By careful formulation, gels can be made to set over the range 0–50°C but this process is limited to relatively soft textures.

1.5.7 Alginate/pectin gels (Toft, 1982; 1986)

When used alone, high-methoxy (HM) pectins are only able to form gels at high sugar solids levels within a narrow pH range. When sodium alginate is included, gel formation takes place at low solids and a wider pH range.

Fruits naturally rich in pectin, such as apples, form gels when a sodium alginate solution is added after cooking. Rigid gels are formed with addition of high-methoxy pectin and sodium alginate high in guluronic acid. Softer gels are formed by addition of high mannuronic alginate. The alginate–pectin synergism is one of very few interactions for alginate with other hydrocolloids, and the only one of commercial value.

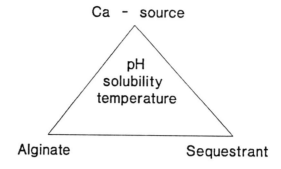

Figure 1.13 Factors influencing gel formation and properties.

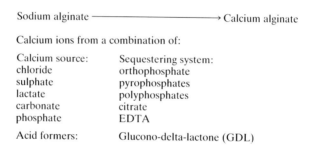

Figure 1.14 Control of gel formation.

1.6 Alginate processes for food production

1.6.1 General

Factors influencing alginate gel formation and properties are illustrated in Figures 1.13 and 1.14. The chemical and physical changes involved are rather complex and knowledge of suitable alginates and calcium release in the chosen sequestering system is required. In fact the alginate grade and the calcium sequestering system (including calcium source and sequestering agent/agents) must be matched with the process developed to manufacture the food.

Calcium release can be studied by the use of an ion selective Ca-electrode (Figure 1.15). All basic and fundamental knowledge of alginates

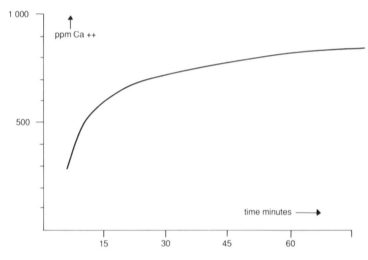

Figure 1.15 Calcium ion release in a 0.6% calcium sulphate, 0.06% tetrasodium pyro-phosphate sequestering system measured by specific calcium ion electrode.

and calcium sequestering systems will certainly help in developing restructured products based on the alginate technology. However, a successful product development of such a restructured food product will always involve trials, especially regarding scale-up, which has to be empirically evaluated.

1.6.2 Appropriate food raw materials

Almost any food material that can be pulped can also be restructured, recombined or reformed. Several benefits may be obtained by restruc-turing a variety of products. Restructuring allows the utilisation of cheap raw materials (fruit or vegetable pulps, meat pieces, abattoir waste, fish mass). It may give a standard automated process in terms of mechanical properties (handling resistance), and a convenient product in terms of shape, reproducibility (identical shape), heat resistance, and freeze–thaw stability.

Recently new developments on meat products have exploited alginate restructuring. The process for preparing alginate/calcium gel structured meat products has been patented (Schmidt and Means, 1986) and there have been several articles published on this subject (Means and Schmidt, 1986; Means et al., 1987; Ernst et al., 1989; Trout, 1989; Trout et al., 1990).

1.6.3 Commercial alginate gel restructured products

Among the restructured products already on the market are the following:

- Onion rings
- Pimiento olive fillings
- Anchovy olive fillings
- Apple pieces for pie fillings
- Cocktail berries
- Meat chunks for pet food
- Shrimp-like fish products
- Fish patties

All the product examples mentioned above have one thing in common; they utilise the specific properties of alginate to obtain shape control. Alginate offers controlled gelling, even at temperatures close to freezing, it gives complete heat resistance and freeze–thaw stability, and it provides the desired mechanical strength. The amount of alginate needed is usually between 1 and 2% of the final product weight. Some recipe examples are given here.

Restructured onion rings

Ingredients

Alginate (fine mesh, high viscosity, high gel strength)	8.2 g
Wheat flour	126.4 g
Salt (NaCl)	1.4 g
Onion powder (16 mesh grade)	159.0 g
Distilled water	705.0 g (ml)
Total	1000.0 g

Setting bath: Calcium chloride, $CaCl_2 \cdot 2H_2O$, 5% solution in water (2.5 l setting bath per 1000.0 g onion mass).

Method
1 Mix the dried onion with 634 ml of the distilled water.
2 Allow to rehydrate for approximately 1 h.
3 Thoroughly blend alginate, flour and salt to a dry mix.
4 When the onion is rehydrated, stir the onion mass into the rest of the water.
5 Add the dry mix of alginate, flour and salt, and stir the onion mass until it is homogeneous (1 min is usually sufficient).
6 Prepare the setting bath.
7 Extrude the onion mass through an annulus, and transfer to the setting bath where restructuring of the onion takes place.

Pimiento filling for olives

Ingredients

Alginate (standard mesh, high viscosity, high gel strength)	18 g

Guar gum	10 g
Pimiento pulp	300 g
Water	672 g (ml)
Total	1000 g

Setting bath

Calcium chloride (CaCl$_2$ · 2H$_2$O)	50 g
Water	950 g (ml)

Ripening brine

Calcium chloride (CaCl$_2$ · 2H$_2$O)	30 g
Sodium chloride	80 g
Lactic acid	8 g (ml)
Water	882 g (ml)

Method

1 Alginate (gel former) and guar gum (thickener and water binder) are dissolved in the water and added to the pimiento pulp.
2 The mix is stirred for 2–3 minutes and extruded onto a conveyor belt.
3 The mix is formed to a sheet approximately 2–3 mm thick. This is done by feeding the pimiento mix between two rolls which are lightly sprayed with calcium chloride to prevent the mix sticking to the rolls.
4 The conveyor belt runs into a setting bath of 5% calcium chloride. Calcium ions react with the alginate molecules and instantaneously form a heat stable gel.
5 The setting time depends on the thickness of the sheet. For a thickness of approximately 2–3 mm this will take 10–15 min.
6 The pimiento sheets are stored in an ageing bath.
7 After storage the sheets are washed to remove excess calcium chloride. They now have a strong and non-brittle texture, allowing them to be cut, sharply bent and rapidly injected into pitted olives.

Anchovy filling for olives

Ingredients

Alginate (standard mesh, high viscosity, high gel strength)	25.0 g
Dicalcium phosphate	10.0 g
Anchovy, minced	290.0 g
Water	675.0 g (ml)
Total	1000.0 g

Method

1 Make a paste with the minced anchovy and 290 ml of the water.
2 Dissolve the alginate in 350 ml of the water by sprinkling the powder gently into the water and stirring vigorously with a mechanical stirrer. Prepare a homogeneous and lump free solution.
3 Add this solution to the anchovy paste while stirring.
4 Dissolve the dicalcium phosphate in the rest of the water (35 ml). Add this to the mix.
5 The anchovy paste is now suitable for machine handling and filling.

Pet food

Ingredients

Alginate (coarse mesh, high viscosity, high gel strength)	8 g
Calcium sulphate	6 g
Tetrasodium pyrophosphate	1 g
Soy oil	20 g
Minced meat, offal or fish as required	600 g
Water	365 g (ml)
Total	1000 g

Method

1 Mince the meat twice, mechanically.
2 Stir the alginate and tetrasodium pyrophosphate into the soy oil.
3 Add the mixture to the minced meat and mix mechanically for 2 min.
4 Slurry the calcium sulphate into the water and add to the minced meat.
5 Mix the whole mixture for another 2 min.
6 If the mixture is to be formed into chunks, this must be done immediately. The time for surface gel formation is approximately 10 min and the time for complete gelation is approximately 20 min.
7 Alternatively the mixture can be left to gel first, and then cut into pieces.
8 The pet food, with or without a gravy or meat paté, is autoclaved at 120°C for 45 min.

1.6.4 Examples of production formats

Different examples of suitable production processes are shown in Figures 1.12, 1.16 and 1.17.

The pet food process line, Figure 1.12, represents the state of the art for industrial production of food based on the alginate gel concept. Here both diffusion and internal gel setting are utilised together for semi-continuous production.

We are still gaining knowledge in the field of alginate gel technology. In the near future, it is likely that process technology like this will be adopted by the food industry for production of foodstuffs for human consumption.

1.7 Thickening and stabilising

1.7.1 Thickening

As described above, alginate develops viscosity in a water solution. In practice, it is difficult to make a sharp distinction between thickening and gelling. A very weak gel may appear as a thick solution, and thickening may often be the result of limited calcium alginate cross-linking.

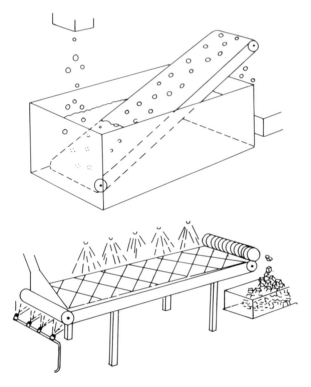

Figure 1.16 Top: Calcium chloride setting bath. Bottom: setting belt with calcium chloride spray.

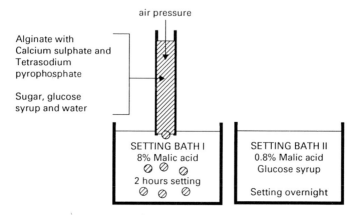

air pressure

Alginate with
Calcium sulphate and
Tetrasodium
pyrophosphate

Sugar, glucose
syrup and water

SETTING BATH I
8% Malic acid

2 hours setting

SETTING BATH II
0.8% Malic acid
Glucose syrup

Setting overnight

Figure 1.17 Laboratory process for production of artificial cherries.

Several applications for alginate exist in bakery products. One important application, which demonstrates alginate gel properties, is its use to thicken cold-prepared bakery creams and at the same time make them heat resistant or bakefast as well as freeze–thaw stable. Additional uses include thickening of pie fillings, reducing the tendency to soak into the pie pastry, and thickening batter to give a short flow which gives good deposition control and pick-up which is easier to handle on production machinery. Finally, the water-binding properties of alginate allow its use as a moisture-retaining additive in pastries.

For sauces, the thickening properties of alginate can be used in two contrasting ways, as a temporary or a delayed thickening agent. An example of the first process is the use of alginates to prevent meat from settling out in gravy prior to filling into cans. When the filled cans are retorted, the heat treatment reduces the thickening effect of alginate, which is no longer required.

The opposite effect can be used to facilitate heat exchange in cans during retorting. The effect is obtained by using a mixture of alginate and a slowly soluble calcium salt. The combination of slow solubility and heat delays gel formation until the heat treatment has taken place. Rapid heat transfer in the low viscosity sauce then allows a shorter heating cycle than with a pregelled product. Food emulsions such as sauces and salad dressings can be thickened and stabilised to prevent phase separation by including alginate in the aqueous phase.

1.7.2 Stabilising

Water-soluble alginates can act as stabilisers in systems consisting of particles or droplets dispersed in water. Such systems include oil-in-water emulsions such as ice cream and salad dressings, and solids suspended in water, such as fruit squash. The alginate acts as a stabiliser to prevent separation by increasing the viscosity of the aqueous phase and by producing charged films at the interfaces, so that the individual particles or droplets tend to repel one another avoiding coalescence and phase separation.

Ice cream was the first application for alginate in the food industry. Alginate addition reduces the size of the ice crystals and produces a smooth product. Alginate also prevents syneresis, and prolongs the melt-down of the ice cream. In ice cream, dairy products and other systems where calcium ions are present, the alginate is usually mixed with a low level of sodium phosphate. This prevents precipitation of calcium alginate and binds free calcium which may hinder or prevent alginate hydration. After the phosphate is combined with calcium, subsequent release of calcium cross-links the alginate leading to thickening and gelation.

1.7.3 Thickening/stabilising with propylene glycol alginate (PGA)

At pH values below or around the pK_a of alginic acid (approximately 3.5), alginate will not be an effective thickener/stabiliser. As the pH decreases, alginate takes up protons and loses net negative charge. Eventually alginic acid starts to precipitate. Propylene glycol alginate (PGA), in contrast, possesses valuable functionality even in low pH conditions. The presence of the lipophilic propylene glycol ester groups provides PGA with emulsifying power and makes it acid tolerant and less calcium reactive than the sodium alginate original. The remaining unesterified acid groups retain some negative charge even down to pH 2.75 and will participate in weak but significant cross-linking inter-actions with calcium or with proteins which, at this low pH, usually carry a net positive charge. These PGA properties are utilised in the stabilisation of milk proteins under acidic conditions, in yogurt and in lactic drinks, for pulp stabilisation in acidic drinks in general, and in the extremely important applications of PGA as stabilisers of foam in beers, lagers and also in low alcohol drinks as these gain market share. In acidic sauces, such as salad dressings, PGA is also appreciated as a particularly effective thickener and stabiliser.

1.8 Film formation

When a thin layer of alginate gel or alginate solution is dried, a film is formed. This film reduces water loss, and can be used in pastries to

Table 1.4 Alginate properties utilised in food products.

Gel forming	Pet food
	Restructured fruit and vegetables
	Restructured fish and meat
	Puddings and desserts
	Cold prepared bakery creams
	Fruit preparations, bakery jam
Thickening/waterbinding	Tomato ketchup, tomato sauce
	Soups, sauces
	Milk shakes
Stabilising	Ice cream
	Mayonnaise
	Whipped cream
	Margarine
	Salad dressing (PGA)
	Fruit juice (PGA)
	Beers, lagers (PGA)
Film forming	Glazes for frozen meat and fish
	Film coatings for fresh meats
	Coatings for cakes and cookies

Table 1.5 Application areas: food.

Restructured foods	Onion rings
	Pimiento fillings for olives
	Fruit fillings
	Fish, meat and poultry
Pet food	Alginate gelled chunks
Bakery products	Bake-stable filling creams
	Bake-stable marmalade
Ice cream	Ice cream
	Sorbet, fruit ice
Jams and marmalades	Yogurt fruit mix
	Diabetic and low-sugar jam
	Luxury marmalade
Dressing and sauces (emulsions)	Salad dressings
	Mayonnaise
	Ketchup, tomato sauce
	Low-calorie margarine/spreads
Desserts and dairy products	Instant desserts
	Jelly, puddings
Toppings, sauces and ripple syrups	
Drinks	Beers, lagers
	Juice, squash
	Liquors
Biotechnology	Immobilisation of bacteria and yeast
Feed (part-processed seaweed)	Fish feed

prevent water passing from the filling into the remainder of the cake, it can be used in cake icings to prevent adhesion to the wrapping and, simultaneously, act as an anti-cracking agent.

Alginate can be used to protect frozen fish from oxidation and loss of water by stabilising the coating layer of ice and making it more impermeable to oxygen.

Meat carcasses and meat pieces can be protected by a calcium alginate film which both reduces water loss and improves the bacteriological quality of those products. The same system can be applied to poultry and to hamburger-like products.

1.9 Summary

Alginate properties utilised in food production are listed in Table 1.4. Alginate application areas, in the food industry, are listed in Table 1.5.

As previously explained the field of alginate gel technology is still growing. The future general demand for more effective utilisation of the world food resources will certainly trigger new developments in this area. Food technology process development will be targeted towards semi-continuous and continuous production away from batch processing. Alginates will be used in many of these developments because alginate gel formation can take place at room temperatures and the sophisticated processes for restructuring products using alginate gel technology are highly developed, for example in pet food chunk production.

The market for prepared foods has shown an increase for many years, and will probably continue to grow with the microwave oven being more widely used and the development and common availability of frozen storage technology. This will create a general demand for hydrocolloids to control the functional properties of such food products, especially for alginate technology due to the alginate gelation and heat stability characteristics.

An alginate gel technology has also been developed for immobilisation and encapsulation and other purposes in biotechnology. Many of the new developments in biotechnology can be utilised in food products. Relevant examples are yeast cells immobilised in alginate beads for beer and ethanol production, as well as in the secondary fermentation of champagne (Onsøyen, 1990).

References

Cottrell, I.W. and Kovacs, P. (1980) In: *Handbook of Water-soluble Gums and Resins*, R.L. Davidson, ed., McGraw-Hill, New York.

Ernst, E.A., Ensor, S.A., Sofos, J.N. and Schmidt, G.R. (1989) Shelf-life of algin/calcium restructured turkey products held under aerobic and anaerobic conditions. *J. Food Sci.*, **54**(5), 1147–1154.

Grant, G.T., Morris, E.R., Rees, D.A., Smith, P.J.C. and Thom, D. (1973) Biological interactions between polysaccharides and divalent cations: the egg-box model. *FEBS Lett.*, **32**, 195.

Grasdalen, H. (1983) High-field ^{1}H-NMR spectroscopy of alginate. Sequential structure and linkage confirmations. *Carbohydr. Res.*, **118**, 255–260.

Means, W.J. and Schmidt, G.R. (1986) Algin/calcium gel as a raw and cooked binder in structured beef steaks. *J. Food Sci.*, **51**(1), 60–65.

Means, W.J., Clarke, A.D., Sofos, J.N. and Schmidt, G.R. (1987) Binding, sensory and storage properties of algin/calcium structured beef steaks. *J. Food Sci.*, **52**(2), 252–262.

Onsøyen, E. (1990) Marine hydrocolloids in biotechnological applications — the topical demand: to what extent must they be purified and characterized. In: *Advances in Fisheries Technology and Biotechnology for Increased Profitability*, papers from the 34th Atlantic Fisheries Technological Conference and Seafood Biotechnology Workshop, M.N. Voigt and J.R. Botta, eds, Technomic Publishing Co. Inc., Lancaster, USA, 265–286.

Protan Biopolymer A/S (1988) *Protan Alginates in Pet Food*.

Protan Biopolymer A/S (1990) *Technical Information — Alginates*.

Rees, D.A. (1969) Proposed gelation mechanism for carrageenans. *Adv. Carbohydr. Chem. Biochem.*, **24**, 267–332.

Schmidt, G.R. and Means, W.J. (1986) Process for preparing algin/calcium gel structured meat products. US Patent 4,603,054.

Smidsrød, O. (1972) Molecular basis for some physical properties of alginates in the gel state. *Faraday Disc. Soc.*, **57**, 263–274.

Toft, K. (1982) Interactions between Pectins and Alginates. *Prog. Food Nutr. Sci.*, **6**, 89–96.

Toft, K., Grasdalen, H. and Smidsrød, O. (1986) Synergistic gelation of alginates and pectins. In: ACS Symposium Series No. 310, *Chemistry and Function of Pectins*, M.L. Fishman and J.J. Jen, eds, American Chemical Society.

Trout, G.R. (1989) Color and bind strength of restructured pork chops: effect of calcium carbonate and sodium alginate concentration. *J. Food Sci.*, **54**(6), 1466–1470.

Trout, G.R., Chen, C.M. and Dale, S. (1990) Effect of calcium carbonate and sodium alginate on the textural characteristics, color and color stability of restructured pork chops. *J. Food Sci.*, **55**(1), 38–42.

2 Carrageenan

W.R. THOMAS

2.1 Introduction

Throughout history red seaweeds have been harvested and used as foods. The Chinese have recorded uses back to 600 BC, while the early Irish used them to form milk puddings. In the Far East they are consumed as salads and gels. The carrageenan extracted from seaweed is not assimilated by the human body, providing only bulk and no nutrition, but it does provide outstanding functional properties which can be used to control moisture and texture and to stabilise food systems.

Carrageenan is a naturally occurring polysaccharide material which fills the voids in the cellulosic plant structure (Figure 2.1). As a result of its water gelling and milk protein interaction, this extracted hydrocolloid material is extensively used by the food industry to gel, thicken and stabilise food systems.

The red seaweeds produce extracts which compose a family of hydro-colloids including agar, furcellaran and three types of carrageenan. All of these have a galatose backbone joined together by alternating glycosidic linkages. However, they are differentiated by the number and position of ester sulphate groups and the amount of 3,6 anhydro-D-galactose which they contain. This results in a wide range of gelling properties, from the very brittle agar gels to carrageenan gels and the non-gelling lambda carrageenans (Figure 2.2). Carrageenans have a wide range of appli-cations because, within the sub-families of carrageenan hydrocolloids, a very broad range of properties can be developed by using the interactions between various types of carrageenans (kappa, iota, lambda), milk proteins and other hydrocolloids (konjac flour, locust bean gum).

2.2 Raw materials

The primary red seaweed sources of carrageenan are *Chondrus crispus*, a small cold water weed producing kappa and lambda types, warm water *Eucheuma* species producing kappa and iota types, and the large cold water *Gigartina* species which also produces the kappa and lambda types (Figure 2.3). With the large growth in seaweed farming areas in the Philippines and Indonesia, the availability of raw materials is assured. In

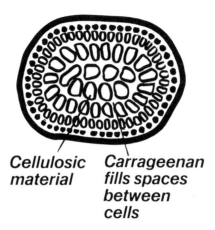

Cellulosic material

Carrageenan fills spaces between cells

Figure 2.1 Cross-section of red seaweed.

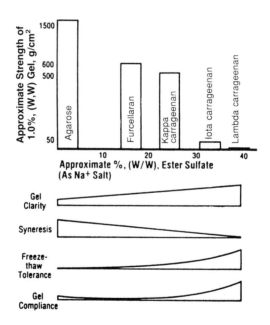

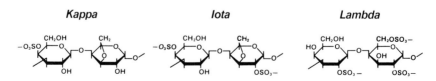

Figure 2.2 Properties of carrageenan gels as a function of the ester sulphate content.

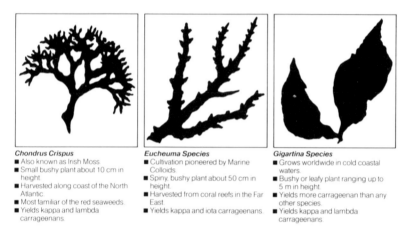

Chondrus Crispus
- Also known as Irish Moss.
- Small bushy plant about 10 cm in height.
- Harvested along coast of the North Atlantic.
- Most familiar of the red seaweeds.
- Yields kappa and lambda carrageenans.

Eucheuma Species
- Cultivation pioneered by Marine Colloids.
- Spiny, bushy plant about 50 cm in height.
- Harvested from coral reefs in the Far East.
- Yields kappa and iota carrageenans.

Gigartina Species
- Grows worldwide in cold coastal waters.
- Bushy or leafy plant ranging up to 5 m in height.
- Yields more carrageenan than any other species.
- Yields kappa and lambda carrageenans.

Figure 2.3 Major raw material sources.

addition, within the last ten years there has been a large increase in world-wide extraction plant capacity. Because this is an agricultural product, production planning is important to assure that materials are available when needed. Cold water seaweeds are harvested once per year while the warm water seaweeds grow on a 3-month cycle but may require another 3 to 6 months before processing is complete.

As with all agricultural products, quality starts in stock selection and proper harvesting. For this reason most manufacturers have field personnel to ensure that the seaweed is harvested at the right time and dried quickly to preserve the quality of the carrageenan contained in it. After it has been dried to the proper moisture content, the seaweed is bailed and shipped to the extract plant where it is warehoused until extraction is required. Because of the very long lead times in obtaining quality weeds and the need to service the ever-changing food industry, the inventory cost for the manufacturers is very high, making proper forecasting essential.

2.3 Manufacturing

At the manufacturing site, the baled seaweed is tested again and various lots are selected to produce the desired end product. Proper selection of the raw materials and an understanding of the influence of the extraction process on the final extract are vital to the production of high quality and consistent product. However, even the best manufacturer will blend

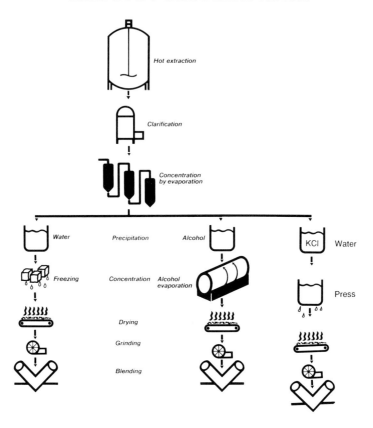

Figure 2.4 Various manufacturing processes.

various finished extracts in order to maintain a consistant quality and provide the properties needed to meet particular customer needs or specifications. A single manufacturer may market 200 to 300 different blends.

After the seaweed has been selected and the appropriate process prescribed, it is further cleaned to remove any sand before being placed in the extraction tanks (Figure 2.4). Here the various seaweeds are coprocessed with varying amounts of alkali to optimise gelling and viscosity characteristics, and break down the cellulose so that the carrageenan can be extracted from the seaweed. After clarification and concentration the carrageenan is precipitated by various methods before drying and grinding.

The extract is again tested for various properties before customised blending for specific applications and customer needs is completed.

Table 2.1 Carrageenan regulatory approvals.

1 *Food and Drug Administration (FDA)*	
Established as Generally Recognized as Safe (GRAS)	21 CFR 182.7255
Approved as food additive	21 CFR 172.620
2 *Food Chemicals Codex*	
Establishes food grade specifications	3rd Edition National Academy Press, 1981
3 *US Pharmacopeia/National Formulary*	
Establishes pharmaceutical grade specifications	US Pharmacopeia Convention, 1985
4 *European Community (EC)*	
Established European Food Grade specificatons	Official EC Journal L 223, Volume 21, 8/14/78
5 *Food Agriculture Organization/World Health Organization Joint Expert Committee on Food Additives*	
Confirms safety of carrageenan and sets international food specifications	Evaluation of Certain Food Additives and Contaminants World Health Organization Technical Report 710, 1984

2.4 Regulation

Carrageenan has been safely used for centuries, reviewed by regulatory agencies and tested using standard biological and toxicological methods. Carrageenan is accepted world-wide with very few restrictions (Table 2.1).

Carrageenan is defined by the United States Department of Agriculture (USDA) food and drug administration and food chemicals codex as a purified seaplant extract having a molecular weight greater than 100 000 and possessing useful thickening and gelling properties.

2.5 Functional properties

Carrageenan is a high molecular weight linear polysaccaride made up of repeating galactose units and 3,6-anhydrogalactose (3,6-AG), both sulphated and non-sulphated, joined by alternating α-(1,3) and β-(1,4) glycosidic linkages. Various naturally occurring arrangements of components create three basic types of carrageenan, commonly referred to as kappa (κ), iota (ι) and lambda (λ) carrageenan (Table 2.2). Variations of these components influence gel strengths, texture, solubility, synergisms and melting temperatures of the carrageenan. These variations are controlled and created by seaweed selection, processing and blending of extracts.

As can be observed in Figure 2.5, the thickening and gelling mechanisms of these three components are quite different. For example, kappa

Table 2.2 Major carrageenan extracts.

Properties of carrageenan	Kappa	Iota	Lambda
Solubility			
80°C water	Yes	Yes	Yes
20°C water	Na$^+$ salt soluble K$^+$, Ca^{2+} and NH$_4^+$ salts swell	Na$^+$ salt soluble Ca^{2+} salt swells to form thixotropic dispersion	Yes
80°C milk	Yes	Yes	Yes
20°C milk	No	No	Thickens
50% sugar solution	Hot	No	Yes
10% salt solution	No	Hot	Hot
Gelation			
Strongest gels	With K$^+$ ion	With Ca^{2+} ion	No gel
Gel texture	Brittle	Elastic	No gel
Re-gelation after shear	No	Yes	No
Syneresis	Yes	No	No
Freeze-thaw stability	No	Yes	Yes
Synergism with locust bean gum	Yes	No	No
Synergism with konjac flour	Yes	No	No
Synergism with starch	No	Yes	No
*Acid stability**			
Gels pH > 3.5	Stable	Stable	Stable
Salt tolerance	Poor	Good	Good

* Hydrolysis in low pH systems accelerated by heat

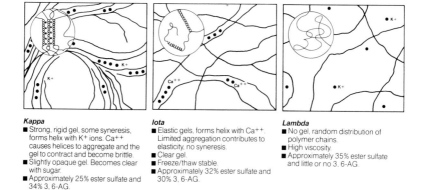

Kappa
- Strong, rigid gel, some syneresis, forms helix with K+ ions. Ca++ causes helices to aggregate and the gel to contract and become brittle.
- Slightly opaque gel. Becomes clear with sugar.
- Approximately 25% ester sulfate and 34% 3, 6-AG.

Iota
- Elastic gels, forms helix with Ca++. Limited aggregation contributes to elasticity, no syneresis.
- Clear gel.
- Freeze/thaw stable.
- Approximately 32% ester sulfate and 30% 3, 6-AG.

Lambda
- No gel, random distribution of polymer chains.
- High viscosity.
- Approximately 35% ester sulfate and little or no 3, 6-AG.

Figure 2.5 Carrageenan gelling mechanisms.

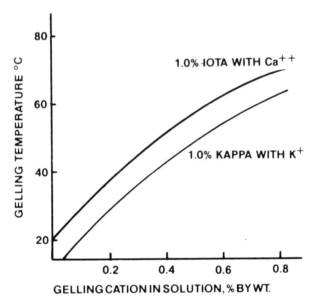

Figure 2.6 Effect of gelling cation on melting and gelling temperatures.

carrageenan forms a firm gel in the presence of K^+, while iota and lambda are only slightly affected. In most cases lambda is used with kappa in milk systems in order to obtain a suspension or creamy gel. Application of these combinations requires experience and understanding of carrageenans. This expertise is readily and willingly provided as a free service by all major manufacturers of carrageenan.

Knowing the ionic content of the system which is being used is important. For example, potassium and calcium are essential for gelation (Figure 2.6); they will also raise the melting and gelling temperatures. All carrageenans are soluble in hot water. Except for lambda, only the sodium salts of iota and kappa carrageenans are soluble in cold water.

The influence of temperature is also an important factor in deciding which type of carrageenan should be used in a food system. All carrageenans are soluble at high temperatures developing very low fluid-processing viscosities. However, kappas and iotas set to various gel textures from 40 to 70°C depending on the cations present. These gels are, of course, stable at room temperature but can be remelted by heating to about 5–10 degrees above the gelling temperature. On cooling the system will re-gel (Figures 2.7 and 2.8).

Once it has been cooled below the gelling temperature, carrageenan is totally stable at pHs normally incurred in food systems. However, it will lose viscosity in systems below pH 4.3 if held at an elevated temperature.

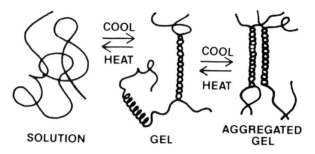

Figure 2.7 Gelation mechanism.

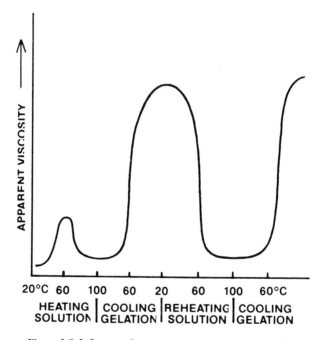

Figure 2.8 Influence of temperatures on carrageenan gels.

In a low pH vat (batch) process, the carrageenan should be added at the last moment after the food has been heated. At moderately low pH, the gelled states are stable but the sols are not, especially when they are subjected to high temperatures for extended periods of time. The result is hydrolysis of the carrageenan which leads to a lower gel strength.

Table 2.3 shows approximate processing times at various pHs for a gel produced with 0.5% kappa carrageenan and 0.2% KCl, such that no

Table 2.3 Gel processing times.

Temperature (°C)	Final pH						
	3	3.5	4	4.5	5	5.5	6
120	2 s	6 s	20 s	1 min	3 min	10 min	30 min
110	6 s	20 s	1 min	3 min	10 min	30 min	1.5 h
100	20 s	1 min	3 min	10 min	30 min	1.5 h	5.0 h
90	1 min	3 min	10 min	30 min	1.5 h	5.0 h	15.0 h
80	3 min	10 min	30 min	1.5 h	5.0 h	15.0 h	2.0 days
70	10 min	30 min	1.5 h	5.0 h	15.0 h	2.0 days	6.0 days
60	30 min	1.5 h	5.0 h	15.0 h	2.0 days	6.0 days	20.0 days
50	1.5 h	5.0 h	15.0 h	2.0 days	6.0 days	20.0 days	60.0 days
40	5.0 h	15.0 h	2.0 days	6.0 days	20.0 days	60.0 days	200.0 days

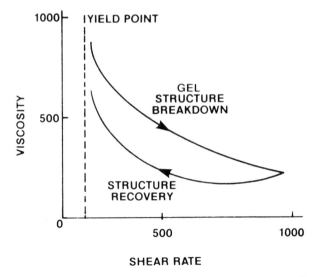

Figure 2.9 Thixotropic properties of dilute iota carrageenan gels.

more than 20–25% of the original gel strength is lost when the sol is cooled. In general, each 0.5 pH unit reduction will decrease potential processing time by a factor of three. Times will vary somewhat, depending on carrageenan concentrations or system ingredients such as salts and sugars. In a continuous process, the processing time should be kept to a minimum (Table 2.3). In systems above pH 4.5 the process conditions become irrelevant.

As mentioned above, kappa gels make a firm, brittle gel which is

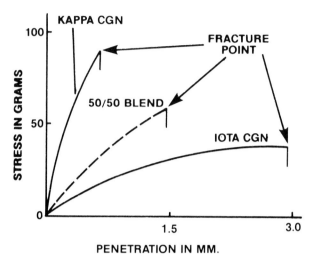

Figure 2.10 Gel texture.

not at all freeze–thaw stable whereas iota systems form a thixotropic (Figure 2.9) or very elastic gel that is freeze–thaw stable. These kappa carrageenans can be blended with iota carrageenans (Figure 2.10) or lambda carrageenans to provide many different textures.

2.6 Synergism with food ingredients

Kappa carrageenan is usually used with locust bean gum in clear water dessert gels to provide a more gelatin-like texture and to decrease syneresis. In addition the combination significantly decreases the cost of making a firm gel (Figure 2.11).

Konjac flour interacts even more strongly with kappa carrageenan to form strong elastic gels which are at least four times the gel strength of kappa alone. It can also be used to form gels which, after heating and cooling, are heat stable above the boiling point (Figure 2.12).

Probably the best known of the synergistic carrageenan interactions is that involving milk proteins. Some of the first uses for carrageenan were in milk gels (flans), the stabilisation of evaporated milk and ice cream mixes where the synergism allows for use levels as low as 0.03%. In these applications the kappa carrageenan not only forms a weak gel (made of double helices) in the water portion of the milk system but it also interacts with the surface area of the casein micelles (Figure 2.13).

Iota carrageenan, in combination with starch, also produces a body which is equivalent to four times that expected by starch alone.

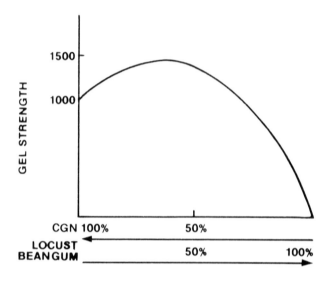

Figure 2.11 Kappa carrageenan/locust bean gum synergism.

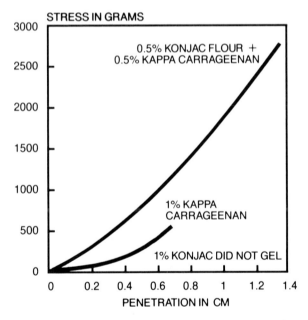

Figure 2.12 Konjac flour and kappa carrageenan synergism.

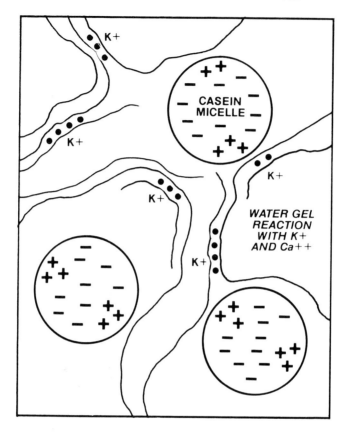

Figure 2.13 Carrageenan synergy with milk protein.

2.7 Food applications

Carrageenan consists of a family of hydrocolloids which have different properties and therefore it has a very wide variety of uses. One of the most important uses is in water-based systems (Table 2.4) which include everything from cake glazes to kamaboko. Some other examples of properties which are important to food formulators are:

- Gel clarity and the high gelling temperatures: important in cake glazes and water dessert gels
- Firm, quick setting gel: valued in processed cheese systems
- Ability to adjust the texture and the melting point: provides the texture needed when carrageenans are used to replace fat in ground meats
- Ability to retard hydration with cations until after the product is

Table 2.4 Carrageenan gels in water-based food systems.

Application	Use level
(a) *Hot processed gelling applications*	
Aquaculture diets	1.00–2.00%
Cake glaze	0.60–0.70%
Cheese, imitation block	2.20–2.70%
Cheese, imitation spread	0.30–0.35%
Desserts, water gels (dry mix)	0.50–0.80%
Desserts, water gels (RTE)	0.60–1.00%
Desserts, water gels (sugar free)	0.60–0.80%
Entrapment/encapsulation	1.00–2.00%
Fabricated/formed foods	2.00–3.00%
Fish gels	0.50–1.00%
Fruit-in-gel	0.80–1.20%
Ham, further processed	0.30–0.50%
Jelly, imitation (low-calorie) dry mix	1.50–2.00%
Jelly, imitation (low-calorie RTE)	1.00–1.50%
Mayonnaise, imitation	0.50–1.00%
Pasta	0.10–0.50%
Peanut butter, extended	2.00–2.20%
Pet food	0.20–1.00% + guar
Pet food, gravy	0.10–0.20% + guar
Pie filling	0.30–0.50%
Poultry nuggets	0.40–0.70%
Poultry, further processed	0.25–0.50%
Red meats, further processed	0.25–0.50%
Seafood, further processed	1.25–1.75% + starch
Sorbet	0.15–0.30% + pectin
Sour cream	0.10–0.20% + LBG
Surimi/kamaboko	0.20–0.30% + starch
Tomato sauces	0.10–0.20% + starch
Whipped cream	0.01–0.05%
(b) *Hot processed thickening applications*	
Anti-icer	0.10–1.00%
Batter mixes	0.10–0.30%
Coffee creamer	0.10–0.20%
Desserts, creamy whipped	0.15–0.30%
Facial mask	1.00–1.50%
Floor polish	1.50–2.00%
Fruit topping	0.30–0.50%
Mayonnaise, imitation	0.40–0.60 + starch
Moisture barriers/meat glazes	0.80–1.20%
Pancake syrup	0.10–0.30%
Salad dressing, hot process	0.20–0.50%
Skin creams	0.50–1.00%
Toothpaste, cold process	0.60–1.20%
Toothpaste, hot process	0.70–1.00%
Variegates/syrups	
(c) *Cold processed thickener applications*	
Barium suspension	1.00–2.00%
Cheesecake (no bake)	0.60–1.00%
Fruit beverages	0.10–0.20%
Liquid cleanser	0.50–1.00% + guar
Mayonnaise, imitation (cold process)	0.40–0.70% + xanthan
Salad dressing (dry mix)	0.60–1.00%
Salad dressing (cold process)	0.20–0.50%

Table 2.5 Carrageenan gels in milk-based milk protein systems.

Application	Use level
(a) *Hot processed milk thickening applications*	
Beverages, acidified (UHT)	
Calcium fortified milk	0.025–0.040%
Chocolate drink	0.02–0.04%
Chocolate milk (HTST)	0.02–0.04%
Chocolate milk (UHT)	0.02–0.05%
Cottage cheese dressing	0.01–0.05%
Cream cheese	0.05–0.08%
Custard (RTE)	0.08–0.12%
Egg nog	0.05–0.12%
Evaporated milk	0.005–0.020%
Ice cream (hard pack)	0.010–0.015% + guar/CMC/LBG
Ice cream (soft serve)	0.02–0.03% + guar/CMC/LBG
Infant formula	0.02–0.03%
Shakes (RTE)	0.02–0.03%
Sterilised milk	0.01–0.03%
(b) *Hot processed milk gelling applications*	
Custards (dry mix)	0.20–0.30%
Dutch vla	0.20–0.04% + starch
Flans (dry mix)	0.20–0.30%
Flans (RTE)	0.20–0.30%
Flans (Soy)	0.20–0.30%
Pudding (cold fill)	0.20–0.60%
Pumpkin pie	0.45–0.55%
(c) *Cold processed milk thickening applications*	
Beverages, nutritional	0.10–0.15%
Breads	2.00–3.00%
Cheese spreads, sauces	0.50–1.00%
Chocolate beverages, dry mix	0.08–0.12%
Chocolate syrups	0.20–0.40%
Desserts, dry mix	0.15–0.20%
Desserts, aerated (mousse)	0.50–1.00%
Ice cream (dry mix)	0.50–0.80%
Meringue topping	0.15–0.25%

pumped and reaches pasteurising temperatures used in ham and poultry breast to control purge and provide firmness

- Low hot viscosities at elevated temperatures: used in UHT systems
- Milk solids content (Table 2.5): makes systems economical because of the interaction with milk protein
- Control of viscosity: prevents cracking and syneresis when filling pie shells (e.g. pumpkin pie)
- Wide range of texture: flans and puddings require textures ranging from brittle to creamy and from light to heavy bodies depending on where they are marketed.
- Synergism with konjac flour, locust bean gum and starch: allows a

variety of gel textures, melting gels and non-melting gels to be produced

The use levels shown in the applications charts (Tables 2.4 and 2.5) are for commercial products which contain cationic salts to improve gelling and control hydration, and various co-processed carrageenans and blends of carrageenans to provide the correct texture. Standardising agents are used to assure proper quality (strength) and to enhance dispersion in liquids. In is always wise to discuss the potential use of a carrageenan with a technical representative from the carrageenan manufacturer in order to get advice on which product most closely matches the food product needs as well as process requirements.

Acknowledgement

Original and redrawn figures are used with permission from FMC corporation.

References

Anonymous, FMC Corporation, Marine Colloids Division:
    Carrageenan — Introductory Bulletin A-1
    Carrageenan — Quick Reference Guide A-2
    Carrageenan — A Naturally Occurring Stabilizer A-3
    Carrageenan — Technical Bulletin G-24
    Carrageenan — General Carrageenan Application
    Carrageenan — Water Gelling Properties of Carrageenan Technology G-31
    Carrageenan — Hydration of Carrageenan G-52
    Konjac flour — Introductory Bulletin K-1
Glicksman, M. (1983) Red seaweed extracts (agar, carrageenan and furcellarai). In: *Food Hydrocolloids*, Vol. 2, M. Glicksman, ed., CRC Press, Boca Raton, pp. 73–113.
Towle, G.A. (1973) Carrageenan. In: *Industrial Gums*, 2nd edn., R.H. Whistler, ed., Academic Press, New York, pp. 83–114.

3 Cellulose derivatives

D. ZECHER and R. VAN COILLIE

3.1 Introduction

Cellulose is the world's most abundant naturally occurring organic substance. It constitutes about one-third of all vegetable matter in the world and is the main constituent of cell walls of higher plants. Wood contains about 40–50% cellulose, bast fibres such as flax contain about 80–90% cellulose and seed hairs, notably cotton, contain about 85–97% cellulose (Ott, 1946). Cellulose is a linear polymer of β-(1,4)-D-anhydroglucose, as shown in Figure 3.1.

The degree of polymerisation (DP, n), can be 1000–15 000 for native cellulose, depending on its origin (Krassig, 1985), but is typically 200–3200 for purified, commercially available cotton linters and wood pulp. Each anhydroglucose unit (AGU) contains three hydroxyl groups available for reaction. The polymer can have a maximum degree of substitution (DS) of 3, where DS is defined as the average number of substituent groups per AGU.

Cellulose is insoluble in water and is not digested by the human body. The lack of water solubility, despite the presence of hydrophilic hydroxyl groups, is generally attributed to the existence of extensive intra- and intermolecular hydrogen-bonded cystalline domains. For food use, conversion to a water-soluble form is usually required. Disruption of the hydrogen bonds can be accomplished by derivatisation of cellulose, as in the formation of cellulose ethers. When alkylene oxides are used as reactants, new hydroxyl substituent groups that can further react are formed, and chaining out (i.e. the formation of sidechain adducts) is possible. The extent of derivatisation is measured as the molar substitution (MS), where MS is defined as the average number of moles of substituent groups per AGU, and the value can exceed 3.

3.2 Manufacture

Cellulose ethers are prepared by reacting cellulose with caustic to form 'alkali cellulose', which in turn is alkylated or alkoxylated in the presence or absence of inert diluents. Since only somewhat exotic solvent systems like cupriethylenediamine, sulphur dioxide–dimethylsulphoxide–

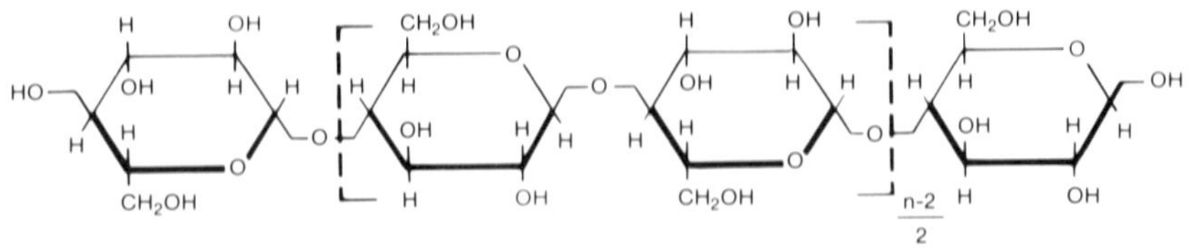

Figure 3.1 Structure of cellulose.

diethylamine (Isogai and Atalla, 1991) and lithium chloride–dimethyl-acetamide (McCormick and Callais, 1987) dissolve cellulose, solution processes are usually cumbersome, impractical and uneconomical. Most commercial derivatisation reactions are conducted *heterogeneously* at elevated temperature in the 50–140°C range under a nitrogen atmosphere, if preservation of high molecular weight is desired (Michie and Neale, 1964). The drawback of this approach is that there is not equal access to all hydroxyls and non-uniform distribution of substituents can result. The caustic used to form alkali cellulose serves to swell the cellulose, disrupt the crystalline regions, promote a more uniform reaction and catalyse alkoxylation reactions. Inert diluents such as acetone or isopropanol serve to disperse the cellulose, provide heat transfer, moderate the reaction kinetics and facilitate recovery of the product.

The two types of reaction typically employed to make cellulose ethers are (a) Williamson etherification involving reaction of alkali cellulose with an organic halide and (b) alkoxylation involving reaction with an epoxide (Donges, 1990), as shown below:

(a) Williamson etherification

$$\text{Cell–OH} + \text{NaOH} + \text{RX} \rightarrow \text{Cell–OR} + \text{NaX} + \text{H}_2\text{O}$$

$$(\text{R} = \text{CH}_3, \text{CH}_2\text{CH}_3, \text{CH}_2\text{COONa, etc.; X} = \text{Cl, Br})$$

(b) Alkoxylation

$$\text{Cell–OH} + \text{CH}_2\text{–CH–R}^* \xrightarrow{\text{NaOH}} \text{Cell–OCH}_2\text{–CH–R}^*$$

$$(\text{R}^* = \text{H, CH}_3, \text{CH}_2\text{OCH}_2\text{CH}_2\text{CH}_2\text{CH}_3, \text{etc.})$$

Note that the Williamson etherification consumes a stoichiometric amount of sodium hydroxide relative to the alkylating agent RX, and generates an equivalent amount of salt NaX, which needs to be removed to make a purified product. The alkoxylation reaction only requires catalytic amounts of caustic; however, to swell the cellulose and make it more accessible to the epoxide, greater than catalytic amounts are used. The caustic is later neutralised, so this too generates salt that needs to be removed to make purified products.

Cellulose derivatives under discussion in this chapter are prepared by reacting alkali cellulose with (a) methyl chloride to form methyl cellulose (MC), (b) propylene oxide to form hydroxypropylcellulose (HPC) or (c) sodium chloroacetate to form sodium carboxymethylcellulose (CMC). Combinations of two or more of these reagents can be used to produce mixed derivatives such as methyl hydroxypropylcellulose (MHPC). Each of these reactions and the commercial grades available are discussed in more detail in later sections of this chapter.

World consumption of MC and derivatives in 1987 was reported to be 61 600 tonnes (135.8 MM lb), including 16 000 tonnes (35.2 MM lb) in the US and 35 400 tonnes (78.1 MM lb) in Western Europe. Growth was forecast to be 3.5% per year between 1987 and 1992. Approximately 1000 tonnes (2 MM lb) was consumed in foods in the US in 1987. Producers include Aqualon Co., Courtaulds Fibres Ltd, Dow Chemical Co., Hoechst AG, Matsumoto Yushi-Seiyaku Co. Ltd, Shin-Etsu Chemical Co. Ltd and Wolff Walsrode AG (SRI International, 1989).

World capacity for HPC was 2300 tonnes (5.1 MM lb) in 1987, including 1800 tonnes (4 MM lb) in the US. A 2% annual growth rate was forecast for the period 1987–1992. The principal HPC producers are Aqualon Co. and Nippon Soda Co. Ltd (SRI International, 1989).

CMC is one of the most widely used of all gums. The world capacity for CMC was 122 800 tonnes (270.7 MM lb) in 1987, including 27 400 tonnes (60.5 MM lb) in the US, 66 500 tonnes (146.6 MM lb) in Europe and 15 900 tonnes (35 MM lb) in Japan. The annual growth rate was forecast to be 2% for the period 1987–1992. Approximately 7000 tonnes (15.5 MM lb) premium-grade CMC was used in food products in the US in 1987, with a 4% annual growth rate forecast for 1987–1992. Major producers of CMC (>4500 tonnes (10 MM lb) per year) include Akzo, Aqualon Co., Billerud AB, Carbose Corp., Daiichi Kogyo Seiyaku Co. Ltd, Daicel Chemical Industries Ltd, Fratelli Lamberti SpA, Hoechst AG and Metsa-Serla (SRI International, 1989).

3.3 MC and MHPC: chemistry and properties

To prepare MC, alkali cellulose is formed by steeping cellulose sheets or chips in caustic, by spraying caustic onto cellulose fibre, by slurrying cellulose in aqueous caustic and pressing out the excess, or by mixing cellulose with aqueous caustic and an inert solvent (Greminger and Krumel, 1980). The alkali cellulose is reacted with methyl chloride in accordance with the Williamson etherification reaction at elevated tem-

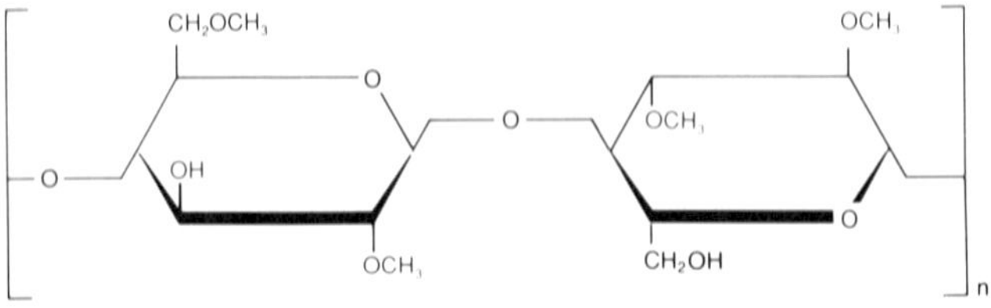

Figure 3.2 Structure of methylcellulose (DS 2.0).

perature, $50-100°C$, and up to $14 \, kg/cm^2$ pressure for several hours (Greminger and Krumel, 1980):

$$Cell-OH + NaOH + CH_3Cl \rightarrow Cell-OCH_3 + NaCl + H_2O$$
$$(MC)$$

Side reactions

$$CH_3Cl + NaOH \rightarrow CH_3OH + NaCl$$

$$CH_3OH + NaOH + CH_3Cl \rightarrow CH_3OCH_3 + NaCl + H_2O$$

Dimethyl sulphate instead of methyl chloride was used under mild conditions, but much by-product formation occurred (Heuser, 1944). Crude reaction products are washed with hot water to remove methanol, dimethyl ether and sodium chloride by-products. A minimum DS of about 1.4 is required for water solubility. At a DS of 2.0–2.2, solubility in organic systems is achieved. Commercial MC products have an average DS ranging from 1.4 to 2.0 (Dow Chemical Co., 1974; Greminger and Krumel, 1980; Aqualon Co., 1989). An idealised structure for 2.0 DS methyl cellulose is shown in Figure 3.2.

These products are white to off-white creamy solids and they are available in various particle sizes, ranging from granular to fine powders. Purity is 98% minimum, with a 2.5% maximum ash, as sulphate. For use in foods, higher purity types are available with sulphated ash below 1.0% and residual heavy metals as specified in the European Pharmacopoeia, the US Pharmacopoeia, European Community Directive 78/663/EEC for MC and the American Food Chemicals Codex. MC is metabolically inert and has a neutral taste and odour.

Preparation of MHPC is similar, with the use of both methyl chloride and propylene oxide as reagents, either sequentially or in combination. Commercial MHPC products have an average M substituent DS of 1.0–2.3 and HP MS of 0.05–1.0 (Dow Chemical Co., 1974; Greminger and Krumel, 1980; Aqualon Co., 1989).

MC and MHPC possess the unique property of being soluble in cold

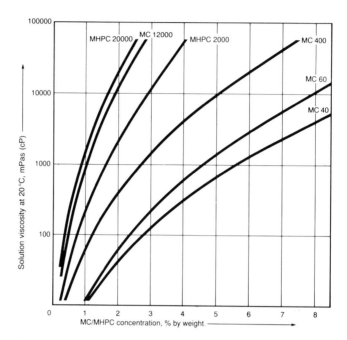

Figure 3.3 The relation between solution viscosity and concentration of various types of MC and MHPC (numbers refer to 2% viscosity).

water but insoluble in hot water. When a solution is heated, a three-dimensional gel structure is formed at a thermal gel temperature ranging from 50 to 90°C. Since MC and MHPC are insoluble in hot water, the crude reaction product can be purified by washing in hot water.

Solutions of MC powder are best prepared by dispersing in a minimum amount of hot water (80–90°C), then adding cold water (0–5°C) or ice to give the final volume and agitating until smooth. MHPC products may require cooling to 20–25°C or below. Solutions of granular MC are best prepared by adding quickly to stirred water to disperse and suspend the polymer. Stirring is continued at a reduced rate until the polymer swells and completely dissolves. MC and MHPC solutions in cold water are smooth, clear and pseudoplastic; their 2% viscosity ranges from 5 to 100 000 mPa s at 20°C and they display little or no thixotropy (Dow Chemical Co., 1974; Greminger and Krumel, 1980; Aqualon Co., 1989). Figure 3.3 shows the relationship between viscosity and concentration of various molecular weight types.

Solution viscosities decrease as temperature increases until the thermal gel point is reached, whereupon the viscosity rises sharply until the flocculation temperature is reached. Above this temperature, the viscosity collapses, as shown schematically in Figure 3.4. Flocculation tempera-

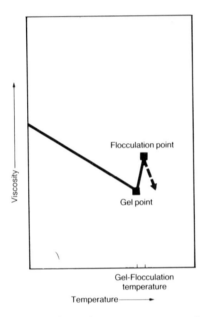

Figure 3.4 Viscosity curve at increasing temperature up to the flocculation temperature.

tures for 0.5% solutions are typically 50–75°C for MC and 60–90°C for MHPC. This phenomenon is due to the weakening of hydrogen bonding between polymer and water molecules and the strengthening of inter-actions between polymer chains. These gels are primarily a result of phase separation and are susceptible to shear thinning. The process is reversible, hence lowering temperature restores the original solution. The thermal gel point is influenced by the type of substitution, DS and solution concentration. The flocculation temperature can be lowered by salts and raised by alcohols such as propylene glycol, as shown in Figures 3.5 and 3.6.

MC and MHPC are stable over a wide pH range of 2–13 in which viscosity is nearly independent of pH. Compatibility with salts depends on the type and concentration of salt and on the type and amount of MC or derivative. Low salt concentrations have little effect on viscosity, but higher levels can 'salt out' the polymer, for example a 2% 7000 mPa s MC is salted out with approximately 7% sodium chloride, 15% potassium chloride or 4% sodium bicarbonate (Aqualon Co., 1989).

Although MC and MHPC exhibit relatively good temperature stability for organic substances, they undergo darkening of colour and slow softening at temperatures above 140°C. At temperatures above 220°C, decomposition occurs. The dry polymers are highly resistant to micro-organisms. However, if MC or MHPC solutions are to be stored for long

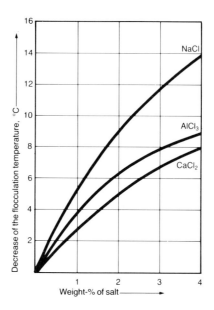

Figure 3.5 Decrease in the flocculation temperature of a 0.5% solution of MHPC (2% viscosity; 600 mPa s) with addition of salt.

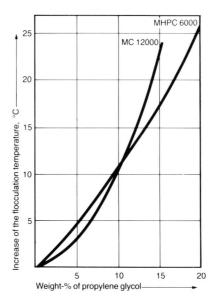

Figure 3.6 Increase in the flocculation temperature of 0.5% solutions of MC/MHPC with the addition of propylene glycol (numbers refer to 2% viscosity).

periods of time, it is recommended that preservatives be used, such as benzoic acid, esters of *p*-hydroxybenzoic acid, sorbic acid and potassium sorbate.

Films with high transparency and tensile strength can be prepared from MC and MHPC. These films are water soluble, but are insoluble in most organic liquids, fats and oils. The oil resistance and thermogelation properties are used to advantage in extruded food products such as seafood and potatoes to reduce oil pick-up during deep-fat frying.

Dilute 0.1% MC and MHPC solutions reduce the surface tension of pure water from $720 \mu N/cm$ to $450-550 \mu N/cm$ at $20°C$. The surface activity and water retention properties of MC in solution justify its use for emulsion-type sauces and in whipped toppings and creams.

3.4 HPC: chemistry and properties

HPC is made by treating cellulose (cotton linters or wood pulp) with aqueous sodium hydroxide, then with propylene oxide in accordance with the alkoxylation reaction. Alkali functions as a swelling agent and catalyst for the etherification. Propylene oxide reacts with water to form poly(propylene glycol) by-products, so the amount of water is minimised to improve reaction yield.

$$Cell-OH + CH_2-CH-CH_3 \xrightarrow{NaOH} Cell-OCH_2CHCH_3 \text{ (HPC)}$$

Side reactions

$$H_2O + CH_2-CH-CH_3 \xrightarrow{NaOH} HOCH_2CHCH_3$$

$$HOCH_2CHCH_3 + CH_2-CH-CH_3 \xrightarrow{NaOH} HOCH_2CHCH_3$$

The reaction may be conducted at $70-100°C$ for $5-20h$ in stirred autoclaves in the presence of organic diluents, neutralised, washed with hot water ($70-90°C$), then dried and ground to provide an off-white, tasteless, granular powder (Klug, 1966; 1967). The purity of food-grade HPC is 99.5% or higher; typically, sulphated ash values are less than 0.2%. Toxicity testing indicates that HPC is physiologically inert.

The typical HP MS of commercial HPC is 3.0–4.5, and usually 3.0–4.0. The idealised structure of a 3.0 MS HPC is shown in Figure 3.7. A wide range of viscosity types are available, ranging from $150 mPa s$ for 10% to $3000 mPa s$ for 1%. HPC is an edible, thermoplastic, non-ionic polymer that is soluble in water below $40°C$ and in many polar

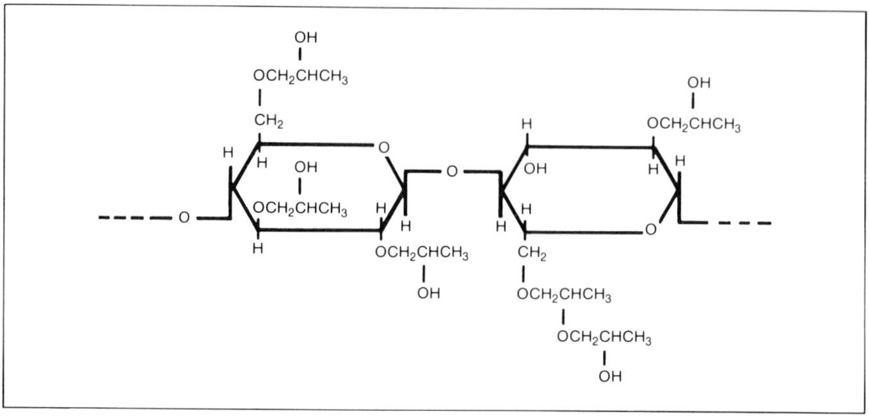

```
                    OH
                    |
                OCH₂CHCH₃                              OH
                    |                                   |
                   CH₂                  H           OCH₂CHCH₃
            H                    O       
              H    OH                  OH        H    H
     --------O                  O                          
                   OCH₂CHCH₃  H            H             O
            H              H        H                  
                   OCH₂CHCH₃      CH₂
                    |             |
                    OH         OCH₂CHCH₃
                                  |
                              OCH₂CHCH₃
                                  |
                                  OH
```

Figure 3.7 Idealised structure of hydroxypropylcellulose (MS 3.0). Toxic substances information: CAS number, 9004-64-2; CAS name, cellulose 2-hydroxypropyl ether.

organic solvents such as methanol, ethanol, propylene glycol and methyl cellosolve. To prepare lump-free, clear aqueous solutions of HPC, it is recommended that the powder is added to hot water or glycerin to form a slurry, which in turn should be added to cold water and agitated until completely dissolved. Alternatively, HPC can be added slowly to the vortex of well-agitated water at room temperature or first dry blended with another product, such as sugar, to help dispersion (Desmarais, 1973).

Aqueous solutions are pseudoplastic, display little or no structure or thixotropy, and have good resistance to shear degradation. Viscosity decreases as temperature is increased, typically by about 50% for every 15°C rise. As the temperature reaches 40–45°C, HPC precipitates from solution (Desmarais, 1973; Aqualon Co., 1987), as shown in Figure 3.8.

The transformation from dissolved to precipitated state occurs without the formation of a gel, contrasting with MC and MHPC. The precipitation phenomenon is completely reversible and the polymer will redissolve below 40°C. The precipitation temperature is lowered by the presence of high concentrations of salts or organic substances such as sucrose that compete for the water in the system (Klug, 1971; Aqualon Co., 1987). For example, the precipitation temperature of 0.5% HPC is lowered to 36°C, 32°C, 20°C and 7°C in the presence of 20%, 30%, 40% and 50% sucrose. Similarly, the precipitation temperature of 1% HPC is lowered from 41°C to 38°C and 30°C in the presence of 1% and 5% sodium chloride respectively, and at high concentrations of dissolved inorganic salts, there is a tendency for HPC to be salted out from solution.

The pH of a 1% HPC solution is typically 5.0–8.8. Since HPC is non-ionic, its viscosity remains unchanged as the pH is varied from 2 to 11.

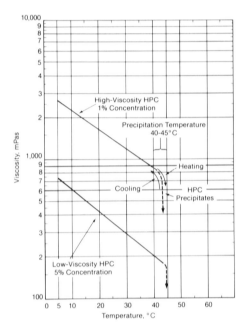

Figure 3.8 Effect of temperature on viscosity of aqueous solutions of hydroxypropyl-cellulose (HPC).

To minimise viscosity loss due to degradation, it is recommended that solutions be stored at pH 6–8.

Aqueous solutions of HPC display surface activity. A 0.1% HPC solution has a surface tension of 440 µN/cm at 25°C, significantly below the value of 741 µN/cm for water (Aqualon Co., 1987). Because of this, HPC can function as an emulsifying agent and whipping aid in whipped toppings and creams (Desmarais, 1973).

HPC solutions are compatible with most natural and synthetic water-soluble polymers including CMC, hydroxyethylcellulose, MC, gelatin, sodium caseinate, polyethylene oxide, guar gum, alginates and locust bean gum. Blending CMC with HPC has a synergistic effect on viscosity since the viscosity of the blended solution is higher than either polymer alone.

Owing to the high level of hydroxypropyl substitution, HPC shows improved resistance to microbiological attack compared with other polymers. However, for prolonged storage of solutions, a preservative is recommended.

HPC is an excellent thermoplastic that can be processed by virtually all fabrication methods, such as injection and compression moulding, blow moulding, injection foam moulding and vacuum forming. It has a softening point in the 100–150°C range. At temperatures in the 250–500°C range it can be completely oxidised (burned off), leaving little residue. It also has excellent film-forming properties that make it a useful material for fabrication of film and sheet, and useful for coatings on paper, food products and pharmaceutical tablets (Aqualon Co., 1987). Films of HPC are characterised by the following outstanding properties:

- Excellent flexibility
- Lack of tackiness
- Good heat sealability
- Ability to act as a barrier to oil and fat.

HPC can be used to modify other coating resins such as shellac, ethylcellulose, CMC and starches. Generally, it improves flexibility and toughness, reduces water resistance and reduces the tendency of the film to crack.

3.5 CMC: chemistry and properties

The manufacture of CMC involves treating cellulose (chemical cotton or wood pulp) with aqueous sodium hydroxide followed by reaction with monochloroacetic acid or sodium monochloroacetate in accordance with the Williamson etherification reaction:

$$Cell-OH + NaOH + ClCH_2COONa \rightarrow$$
$$Cell-OCH_2COONa + NaCl + H_2O$$
$$(CMC)$$

A side reaction produces sodium glycolate:

$$ClCH_2COONa + NaOH \rightarrow HOCH_2COONa + NaCl$$

The use of esters of monochloroacetic acid, such as the isopropyl ester, has also been reported (Taguchi and Ohmiya, 1985). Cellulose sheets can be steeped in alkali, the excess pressed out, shredded and sodium monochloroacetate added before reacting at 50–70°C. Alternatively, a slurry reaction with shredded or cut cellulose can be conducted in an inert, water-miscible diluent such as *t*-butyl alcohol, isopropanol or acetone. At the end of the reaction, excess alkali is neutralised and the crude product is washed with alcohol or acetone–water mixtures that solubilise sodium chloride and sodium glycolate by-products but not CMC (Klug and Tinsley, 1950). There are technical or refined grades that have 94–99% purity. A purified premium grade for use in foods, also known as cellulose gum, has a minimum purity of 99.5%.

Commercial CMCs typically have a DS of 0.4–1.4, but the DS can be

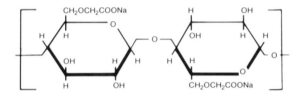

Figure 3.9 Idealised unit structure of carboxymethylcellulose (DS 1.0).

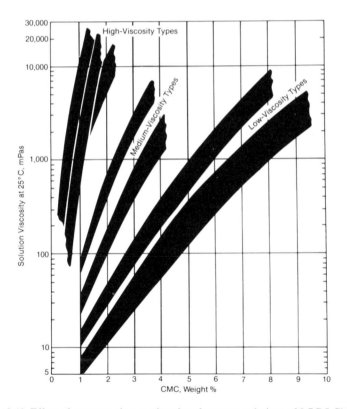

Figure 3.10 Effect of concentration on viscosity of aqueous solutions of 0.7 DS CMC (bands approximate the viscosity range for the types shown).

higher for speciality products. Below about 0.4 DS, CMC is not water soluble. Food-grade CMC typically has a DS of 0.65–0.95. The US Food and Drug Administration defines cellulose gum as the sodium salt of CMC, not less than 99.5% on a dry-weight basis, with a maximum DS of

0.95 and with a minimum 2% viscosity of 25 mPa s at 25°C. Figure 3.9 shows the idealised structure for 1.0 DS CMC.

CMC viscosities can range from 10 mPa s for 2% solutions to 6000 mPa s or more for 1% solutions (Aqualon Co., 1988), as shown in Figure 3.10. Viscosities are controlled by the proper selection of cellulose furnish and/or by oxidative degradation of the crude product with reagents such as hydrogen peroxide to obtain the lower viscosity types. High-viscosity types with 1% viscosities above 1000 mPa s have a DP of up to 3200 and molecular weights up to 700 000. Low-viscosity types with 2% viscosities less than 50 mPa s have a DP of about 400 and molecular weights under 100 000.

The only common solvent for CMC is water, either hot or cold. CMC, a linear anionic polymer, gives non-Newtonian, pseudoplastic solutions, most of which are thixotropic below DS 1.0. CMC solutions are best prepared by direct addition to the vortex of vigorously agitated water, by dry blending with a non-polymeric substance like sugar, by slurrying the CMC in a liquid such as glycerin or propylene glycol, or by using a specially designed mixing device in which CMC is fed through a funnel into a water-jet eductor, where it is dispersed by water flowing at a high velocity (Batdorf and Rossman, 1973).

The degree of substitution and the uniformity of substitution can have a profound effect upon solution properties. As the DS increases, the polymer becomes more soluble as demonstrated by faster hydration rates. The flow characteristics change from a 'structured' or thixotropic flow to a smooth consistency. Similarly, uniformly substituted CMC results in a smoother flowing solution than non-uniform, randomly substituted CMC. When the chain is uniformly substituted, the molecules tend to align under shear and flow smoothly. With randomly substituted CMC, the hydrogen-bonded unsubstituted anhydroglucose units barely swell and tend to associate, creating a three-dimensional network that manifests itself by structured or thixotropic flow. Thixotropy occurs when unsubstituted or crystalline regions in the polymer chain, having been disassociated by shear, reassociate with time to form a three-dimensional network. High-viscosity CMC and low-DS types (0.4–0.7) generally display thixotropy, since these types are less uniformly substituted along the polymer chain. Smooth-flowing CMC types are desirable for food systems such as syrups or frostings where smooth consistency is necessary. Thixotropic CMC finds use in 'grainy' foods such as sauces or purées where suspension and 'short-flow' properties are required (Keller, 1984).

Particle size distribution can also affect CMC performance. Fine-grind CMC is recommended for situations where rapid hydration and viscosity development are important, such as in dry mix beverages. When dispersion or poor mixing conditions are encountered, such as in ice cream, a coarse grind is preferred.

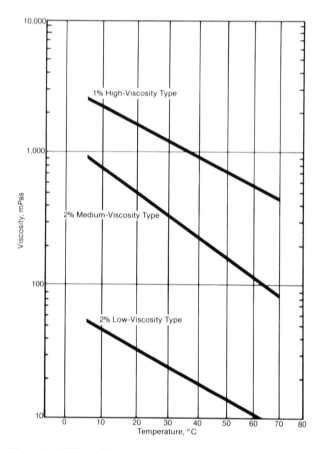

Figure 3.11 Effect of temperature on viscosity of 0.7 DS CMC solutions.

The pH of a 1% CMC solution is typically in the 7.0–8.5 range. The effect of pH on viscosity of CMC across the pH range 5–9 is slight. Below pH 3, the viscosity may increase, and eventually precipitation of the free acid form of CMC may occur. Hence, cellulose gum should not be employed in highly acidic food systems. At pH above 10, a slight decrease in viscosity may occur (Ingram and Jerrard, 1962; Batdorf and Rossman, 1973).

CMC solutions may be degraded by microbiological attack or viscosity-reducing enzymes. Heating to 80°C for 30 min or 100°C for 1 min is often sufficient to destroy most bacteria without affecting the CMC. For prolonged storage, preservatives such as sodium benzoate, sodium propionate or sodium sorbate are recommended. In addition, for maximum stability during storage, the solution pH should be maintained at 7–9 and elevated temperatures, oxygen and sunlight should be avoided.

The viscosity of CMC solutions decreases with increasing temperature. For example, a CMC with a viscosity of 1000 mPa s at 20°C may only have a viscosity of about 100 mPa s at 70°C, as shown in Figure 3.11.

Normally this effect is reversible, but prolonged heating at high temperatures can permanently degrade CMC so that viscosity will not be restored upon cooling.

The presence of salts in solution represses the disaggregation of CMC and therefore affects viscosity. Monovalant cations, with the exception of Ag^+, have little effect when added in moderate concentrations. Divalent metal ions such as Ca^{2+} and Mg^{2+} can lower viscosity, while trivalent metal ions such as Al^{3+}, Cr^{3+} and Fe^{3+} can insolubilise CMC or form gels by complexing with the anionic carboxylic groups (Aqualon Co., 1988). Aluminium CMC gels have not found application in food products because of their astringent taste and slimy mouthfeel (Keller, 1984). The effect of salts on the viscosity of CMC also depends on the order of addition. If CMC is completely dissolved in water and salt is then added, it has only a minor effect on viscosity. If salt is dissolved before CMC is added, the salt inhibits disaggregation (breaking up crystalline regions in the polymer) and lower viscosity results, as shown in Figure 3.12.

CMC is compatible with a wide range of other food ingredients, including protein, sugar, starches and most other water-soluble non-ionic

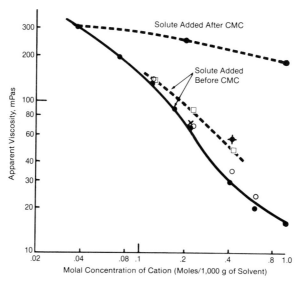

Figure 3.12 Effect of solutes on viscosity of CMC solutions. Solutes used: ●, sodium chloride; ✦, sodium chloride + sodium hydroxide (pH 10.1); ○, sodium sulphate; □, sodium pyrophosphate decahydrate (pH 9.5–9.8); X, potassium chloride.

polymers, over a wide concentration range. In some cases, blending CMC with other polymers produces a synergistic effect on viscosity. For example, when CMC is blended with HPC, the viscosity of the resulting solution is higher than expected from either polymer alone (Aqualon Co., 1988). Similarly, CMC/guar blends are reported to exhibit viscosity synergism in deionised water, but not in 1–3% sodium chloride solutions (Kloow, 1985). CMC is protein reactive and will interact with casein or soya protein. In the isoelectric pH region for casein near 4.5, CMC forms a complex that is water soluble; by itself, casein would precipitate at its isoelectric point. This interaction is the basis for preparing stable acid milk drinks using CMC for its 'protective colloid action' (Keller, 1984).

In contrast to MC, MHPC and HPC discussed above, CMC is not highly surface active. The surface tension of a 1% solution of a 0.7 DS CMC is 710 µN/cm at 25°C compared with 740 µN/cm for pure water.

One of the most important reasons for using CMC in foods is its ability to bind water, thereby preventing syneresis. For example, a small amount of CMC is often used in alginate gels or in starch-based pie fillings to minimise water exudate or syneresis. Furthermore, CMC is a hygroscopic material and will absorb moisture from the air. The amount absorbed depends on the initial moisture content of the CMC, the relative humidity, the temperature and the DS. Higher DS types of CMC are more effective moisture binders. Because of its high water-binding capacity, CMC can be used as a bulking agent in dietetic foods (Sanderson, 1981).

3.6 Major commercial applications in food products

3.6.1 Introduction

Cellulose derivatives (CMC, MC, MHPC and HPC) have been used in food products for over 10 years. At present, their use continues to grow by the development of new applications for existing cellulosic products and by the creation of new food products. The solubility of these cellulose derivatives in water causes the modification of rheological properties and produces structure and texture improvement of food products.

Depending on the conditions of use, cellulosic derivatives can have different functions such as binder, thickener, stabiliser, moisture retention or suspension agent. In some food products only one function is used, whereas in others several functions are used at the same time. Usage levels are generally lower than 1%, and in many cases only 0.1–0.5% is required to obtain the desired technological effect.

Cellulose derivatives (CMC, MC, MHPC and HPC) are used industrially in frozen desserts, soft drinks, bakery products, dairy products, structured and coated products, and other applications.

3.6.2 Frozen desserts

The term 'frozen desserts' includes ice creams, milk ices, sherbets and water ices. In addition to these products, there are some other frozen desserts with different ingredients, such as frozen yogurts.

Ice creams contain about 10% fat, 11% non-fat milk solids, 15% sugar, 0.2–0.3% stabiliser and 0.25–0.5% emulsifier. These concentrations vary considerably from country to country depending on legislation, but the maximum amount of stabiliser allowed is often limited to 0.5%. At present, the stabilisers used in the manufacture of ice creams include CMC, locust bean gum, guar gum, gelatin, alginate and carrageenan. These hydrocolloids are polymers which hydrate more or less rapidly to give viscous solutions. These stabilisers are used in ice creams to increase viscosity, to delay the formation of large ice crystals, to provide a smooth texture and firm bite and to delay melting.

In their study, Moore and Shoemaker (1981) determined the influence of different concentrations of CMC (0–0.2%) on the texture, viscosity and melting time. Ice creams containing no CMC and 0.05% CMC showed significantly higher degrees of iciness than samples containing 0.15% and 0.2% CMC. In addition, the viscosity of ice cream mixes increased with the CMC concentration, and melting speed decreased. Generally, the stabiliser is a blend of two or more hydrocolloids which may show synergistic effects (Keeney, 1982; Bassett, 1988). To make dispersion easier, hydrocolloids are sometimes coated with a mono-glyceride before use. After dispersion, the dissolution of these compounds is easier by mixing the blend at the pasteurisation temperature or the temperature at which the blend is homogenised. Ice creams are then frozen and hardened to $-18°C$ as quickly as possible in order to avoid the formation of large ice crystals. When this temperature is reached, the ice cream is stored at constant temperature. It is possible that temperature will fluctuate during transport and when the product reaches the customer. This means that there is a risk of large crystal formation if the ice cream does not contain stabilisers or is poorly stabilised. Storage tests of ice cream have demonstrated that a high-viscosity CMC and locust bean gum maintain the stand-up, taste and texture of ice creams (Cottrell et al., 1979).

Milk ice generally contains 2–7% fat and a minimum of 11% non-fat milk solids. The concentration of ingredients depends on regulations similar to those for ice cream. While the total solids content is lower than that of ice cream, the content of non-fat milk solids, sweetener and stabiliser is higher.

Formula for hard milk ice
Fat	3%
Non-fat milk solids	13.5%

Sugars	17.5%
Stabiliser	0.4%
Emulsifier	0.15%
Total solids	34.55%

Add water to 100%

Compared with an ice cream, milk ice is not as smooth and does not have as much body and chewiness. To reduce this difference, it is possible to use blends of sucrose and corn syrups. In addition, the concentration of stabiliser will be higher in order to improve texture and body. In this type of product, CMC, as well as guar gum and locust bean gum, tends to cause whey separation. To avoid this, these gums are used in combination with 0.01–0.02% carrageenan. In milk ices, the CMC concentration is about 0.1–0.2%. CMC is used in combination with carrageenan and other gums (locust bean gum, guar gum) in order to obtain the desired texture and body.

The production process is similar to that of ice cream. In the milk ices category, soft ices can be included: their composition is very similar except that they contain less sugar. This results in an increase in the stiffness and dryness when the mix is extruded from the freezer.

CMC, in combination with other stabilisers, is also recommended for stabilising a soft-serve frozen yogurt (Huber and Rowley, 1988).

Sherbet is a sweet frozen dessert which contains fruit (orange, lemon and strawberry) added in the form of juice or pulp. Some countries also allow the addition of small quantities of milk solids.

Formula for sherbet

Milk solids	5%
Sugars	31%
Fruit pulp	15%
Stabiliser	0.4%
Citric acid	0.4%
Total solids	51.8%

Add water to 100%

As with ice cream, CMC stabilises the sherbet's texture, but often is used in combination with other gums such as pectins, locust bean gum and gelatin. The CMC concentration can vary from 0.15 to 0.3% depending on the CMC type used (medium or high viscosity).

In contrast to sherbet, water ice does not contain fruit pulp. In this case, flavours are added to obtain the desired taste. Again, CMC is used in combination with other gums in order to obtain good body and texture. It is also possible to prepare a soft frozen water ice with a overrun of between 25 and 70% from a powder mix containing 0.05–0.5% stabiliser (Huber *et al.*, 1989).

In ripple syrups the amount of sugar is very high, and CMC addition of 0.75–1% may be included to obtain a clear syrup with the desired consistency during freezing and consumption.

3.6.3 Fruit juice-based drinks and products

The classification of fruit juice, or flavour-based drinks and products, can include many products such as concentrated fruit juices, squashes, ready-to-drink beverages, sodas and low-calorie drinks, as well as powder mix preparations. In some of these products fruit juice containing pulp may form a deposit after a period of long storage. It is often difficult to maintain fruit juice pulp in suspension during storage or in dispersion during the drink preparation. To avoid this separation the addition of CMC or a blend of hydrocolloids is often recommended in order to maintain solid materials in suspension. The CMC concentration necessary to obtain good stability depends on the content of soluble solids. When this is high, the CMC concentration is relatively low, because the viscosity of the product is high already. In addition to pulp stabilisation, CMC reduces or prevents the formation of an oily ring in the neck of bottles. When CMC is used in a preparation, it is advisable to hydrate it first in water, at a concentration of about 1%. This solution is then added to the sugar solution already containing the base or the fruit concentrate. It also is possible to add the preservative, colouring and flavour before the CMC solution and then add citric acid or another acid to adjust pH.

In some cases, CMC is used in combination with other gums, such as xanthan gum and propylene glycol alginate (Jackman, 1979; Deleon and Boak, 1984). In this latter case, CMC of a low-viscosity type (25–50 mPa s at 2%) is preferable. When CMC is used alone, medium- and high-viscosity types are recommended, and their addition rate can vary from 0.1 to 0.4% depending on the amount of soluble solids contained in the product and on the degree of dilution before consumption. In addition to its stabilising power, CMC provides body to drinks with a low content of soluble solids as well as to low-calorie drinks.

In ready-to-drink products, the CMC concentration is low (0.025–0.1% depending on the CMC grade) in order to obtain a drink of an acceptable consistency for consumption. In instant fruit and instant breakfast drinks, the addition of CMC, because of its rapid solubility in water, develops the body and mouthfeel required. CMC is also recommended in dry mixes for acid milk drinks (Sirett et al., 1981).

For the preparation of effervescent drinks, such as those prepared by the sodium bicarbonate and citric acid reaction, it is possible to delay this reaction by coating the reagents with a CMC film. These coated reagents are put in water then frozen rapidly in order to avoid interaction. During subsequent thawing interaction occurs in the presence of water to form a carbonic anhydride and produce the drink effervescence (Valbonesi and Cochin, 1981).

For the preparation of instant hot drinks, it was discovered that the incorporation of a suspending agent such as HPC, in combination with a warm liquid (water or alcohol), gives a uniform flavour release over a long period of use (Marmo and Rocco, 1982). In addition to the different products already mentioned, cellulosic derivatives are also used to thicken alcohols for cocktails or for use as flavours in other food products.

3.6.4 Bakery products

Cellulosic derivatives (CMC, MC and MHPC) have been used in the preparation of bakery and pastry products such as biscuits and other similar products for many years.

CMC or MC/MHPC addition may overcome problems with, for example, flour quality, but more often they improve the characteristics of the final products. Generally, the amount used is low and may even reduce the cost of raw materials. CMC is often used in cake mixes containing shortening because of its ability to bind batters (Bayfield, 1962). Cakes of this type are easier to prepare when they contain CMC. Generally, the batter for this type of cake is prepared in one operation by addition of the liquid to the blend of powdered ingredients. The function of CMC and other cellulose ethers is rapidly to develop viscosity, which makes the incorporation of air easier, improving the consistency. During cooking the paste tends to lose its viscosity and increases in volume. With too viscous or too fluid a paste, the cake volume is lower and the crumb structure is less aerated.

The use of CMC in cakes also improves the suspension of raisins, crystallised fruits and chocolate pieces, giving a more homogeneous structure. CMC addition is between 0.1 and 0.5% based on dry ingredients (sugar, flour, shortening, milk solids, dextrose, salt and baking powder). This CMC addition increases the quantity of moisture retained in cakes by about 10% during storage of 4–5 days. This improves the texture and tenderness, factors that affect mouthfeel and eating qualities. These improvements become noticeable after 3–5 days and are the result of better water retention in the crumb, resulting in a less stale product.

In long-life cakes CMC is used to adjust the paste consistency when grains, chocolate sticks or raisins, which tend to fall into the bottom while baking, are used. In addition, CMC improves the water retention, which improves softness. Compared with CMC, MC derivatives tend to lose moisture during baking and form a gel. This gelation leads to some advantages in bakery products, for example in gluten-free breads (Anon., 1985).

The absence of gluten causes an increase in the dough mixing time and a reduction in loaf volume; moreover, the crumb texture is rough and unappealing. The addition of MHPC (0.1–0.5%) to a low-gluten flour can increase loaf volume and improve crumb texture and loaf weight consistency. Other studies on gluten-free breads made from rice flour and potato starch have noted that a combination of CMC and MHPC gives breads with characteristics similar to those of breads made with wheat flour (Ylimaki et al., 1988). In a special bread for slimming, the substitution of some of the wheat flour with blends of bran, CMC and guar gum, in order to produce high-fibre breads, gives good results (Foda et al., 1987).

CMC and MC derivatives can also be used in low-calorie cakes, in protein-rich biscuit formulations (Glicksman et al., 1985) and in ice cream cones. CMC is also used in pastry fillings, fruit pie fillings and icings to improve water retention. In doughnuts, MHPC addition reduces oil absorption during frying and, at the same time, improves the texture. The concentration of cellulosic derivatives in these products is 0.1–0.5%.

3.6.5 Dairy products

Products made with milk or milk derivatives are numerous, but it is possible to classify them into two categories by their pH level. Some of these products have an acid pH, generally lower than 5, obtained by acidification or fermentation. The second category of products has a pH close to that of milk, i.e. 6.7. CMC, being an anionic polymer, reacts with proteins at the isoelectric pH to form a soluble complex. This reaction with milk proteins also occurs with soya and gelatin proteins (Ganz, 1974). The formation of the soluble complex is influenced by the pH, the molecular weight, the CMC concentration and the salt content. Interaction between CMC and casein and other milk proteins gives a soluble complex stable to heat treatment and to storage. Hence, it is possible to prepare pasteurised or sterilised drinks based on yogurt at a pH of 4.4 by adding about 0.5% CMC after fermentation. Drinks based on buttermilk or whey, or prepared by mixing milk and fruit juice, are also stabilised by CMC addition (Shenkenberg et al., 1971).

In general, CMC can be dissolved directly into the acid milk product, or, alternatively, into the milk immediately before adding the acid or fruit juice. In such drinks, the CMC concentration varies from 0.2 to 0.5%. After adding CMC, the product is pasteurised or sterilised before being homogenised, cooled, packaged and stored.

Because of its anionic character, CMC precipitates whey proteins at a pH of 3.2, which can then be used, for example, for cheese preparation (Hansen et al., 1971; Abdel-Baky et al., 1981) to improve the output, texture and body. At neutral pH, CMC also reacts with milk proteins and causes whey separation in low-viscosity mixes such as ice cream and milk ices. It is well known that this separation can be avoided by the addition of carrageenan at a concentration of 10–20% of the CMC used. Carrageenan interacts with the calcium caseinate, preventing the reaction with CMC and avoiding whey separation. In the preparation of desserts (dessert cream, jellified milks), CMC in combination with starches and carrageenan improves texture and avoids syneresis. In acidic whipped creams, CMC and MHPC are used as bulking agents (Hood, 1981). For whipped cream preparations based on vegetable and milk proteins, HPC is used at a concentration of 0.2% (Luzietti and Coacci, 1990).

3.6.6 Structured, extruded and coated products

Structured, extruded and coated products are being requested more and more by the consumer, because they are prepared rapidly at home. These are made from meats, fishes, vegetables, etc., and are generally ready to be consumed directly after cooking.

The development of structured and extruded products allows the creation of new products by using, for example, small pieces of meat or fish, or potatoes too small for French fry preparation. These different ingredients are used to prepare meat sticks (Bernal and Stanley, 1989), fish (da Ponte et al., 1987) and many potato-based products (e.g. croquettes and waffles) by adding 0.5–1% CMC, MC or MHPC. CMC and MHPC have several functions. Firstly, they bind the pieces of meat or fish, making them easier to process as sticks. The cohesion results from the CMC or MC and also from their reactivity with proteins contained in meat or fish.

Structured or extruded products are often coated with a batter and breadcrumbs, which can contain a cellulosic derivative. This improves adhesion of batter to the stick, gives stability to freezing and thawing, and reduces the oil absorption during deep-fat frying. Some potato-based products are coated with a batter containing MHPC in order to reduce oil absorption during prefrying, prior to packing and storage. In addition, MHPC can be used to prevent croquettes from bursting during deep-fat frying prior to consumption.

It also is common to coat non-structured products with a predust, batter and breadcrumb. These meat-, fish- or vegetable-based products are ready to cook or to reheat. Sometimes, to improve the coating adherence, a cellulosic derivative such as CMC can be added at 1–4% to the batter (Suderman et al., 1981).

MHPC is used in the preparation of precooked products, in order to obtain a crispy product after cooking in the microwave oven (D'Amico et al., 1989). The predust should preferably contain at least 20% MHPC having 27–31% (w/w) methoxyl groups and 6–12% (w/w) hydroxypropyl groups.

Since cellulosic derivatives reduce oil absorption, they also are being studied for forming edible barriers in composite foods. These films can contain methylcellulose or HPC ethers mixed with other ingredients such as fatty acid- (Kester and Fennema, 1989) or shellac-based resins (Seaborne and Ebgerg, 1989). These combinations are particularly interesting in composite foods which contain products with very different water contents, for example ice cream cone and ice cream, tomato paste and pizza base, etc.

CMC has a very interesting property which is evident in sausage casings, because it improves peelability when the casings are used in high-speed peeling machines (Higgins and Madsen, 1980).

3.6.7 Miscellaneous food applications

Cellulose derivatives (CMC, MC and MHPC) thicken emulsified and non-emulsified sauces and so stabilise these products by avoiding the formation of an aqueous phase (Vincent and Harrison, 1987). Frequently, cellulosic derivatives are used in blends with other gums such as guar gum, locust bean gum and xanthan gum to obtain the unctuous texture of dressings.

Sauces consumed hot often tend to become fluid during heating. To avoid this viscosity loss, MC and MHPC, because of their thermogelation behaviour, preserve the consistency and appearance of these sauces. This property of MC and MHPC is also useful in soup preparation to obtain consistency and, in general, when a reduction of the concentration of starch is desired.

In instant sauces and soups, cellulosic derivatives rapidly thicken while cooling.

It is recommended that products which contain MC or MHPC and are consumed hot should be prepared cool and then heated, because a direct hot preparation may prevent hydration of MC or MHPC. Depending on the products made and the types of CMC or MC/MHPC used, the concentration used can vary from 0.25 to 1%.

Many other applications exist for cellulose derivatives, in which the film-forming properties of HPC can be used for coating of products sensitive to humidity or the water absorption properties of CMC can be used to resolve problems with exuded fluids from food products (such as meat and poultry) (Midkiff et al., 1990). In pet foods, CMC is recommended as a binder and water retention aid in semi-humid products, and for coating pellets with a high-viscosity grade to give an instant thick gravy with water.

3.6.8 Future developments

In the 1980s, the development of low-calorie products began, and today many consumers are trying to reduce their calorie intake. This reduction in calories is often accomplished by a decrease in the fat and sugar content, which necessitates modification of product formulations. Such modifications can be made only by using new ingredients and additives in order to obtain characteristics similar to those of standard products. Currently, many articles are being published on the use of cellulose derivatives in low-calorie products such as bakery, meat and dairy products.

In addition to low-calorie products, we believe that the next few years will see the further development of convenience foods and microwaveable products, in which cellulose derivatives are already being used to improve some characteristics.

References

Abdel-Baky, A.A., El-Fak, A.M., Abo El-Ela, W.M. and Farad, A.A. (1981) Fortification of Domiati cheese milk with whey proteins/carboxymethylcellulose complex. *Dairy Ind. Int.*, **46(9)**, 29, 31.

Anon. (1985) Methylcellulose in low gluten bread. *Food Feed Chem.*, **17(11)**, 576.

Aqualon Co. (1987) *Klucel® HPC, Physical and Chemical Properties*, Wilmington, DE.

Aqualon Co. (1988) *Aqualon™ Cellulose Gum, Physical and Chemical Properties*, Wilmington, DE.

Aqualon Co. (1989) *Culminal® MC, MHEC, MHPC, Physical and Chemical Properties*, Wilmington, DE; *Benecel high purity MC, MHEC, MHPC, Physical and Chemical Properties*, Wilmington, DE.

Bassett, H. (1988) Stabilization and emulsification of frozen desserts. *Dairy Field*, **171(5)**, 22–25.

Batdorf, J.B. and Rossman, J.M. (1973) Sodium carboxymethylcellulose (Chapter XXXI). In: *Industrial Gums*, 2nd edn, R.L. Whistler, ed., Academic Press, New York.

Bayfield, E.G. (1962) Improving white layer cake quality by adding CMC. *Bakers Digest*, **36(2)**, 50–52, 54.

Bernal, V.M. and Stanley, D.W. (1989) Technical note: methylcellulose as a binder for reformed beef. *Int. J. Food Sci. Tech.*, **24**, 461–464.

Cottrell, J.I.L., Pass, G. and Phillips, G.O. (1979) Assessment of polysaccharides as ice cream stabilizers. *J. Sci. Food Agr.*, **30**, 1085–1088.

D'Amico, L.R., Waring, S.E. and Lenchin, J.M. (1989) US Patent 4,842,874.

Deleon, J.R. and Boak, M.G. (1984) US Patent 4,433,000.

Desmarais, A.J. (1973) Hydroxyalkyl derivatives of cellulose (Chapter XXIX). In: *Industrial Gums*, 2nd edn, R.L. Whistler, ed., Academic Press, New York.

Donges, R. (1990) Nonionic cellulose ethers. *Br. Poly. J.*, **23**, 315–326.

Dow Chemical Co. (1974) *Handbook on Methocel® Cellulose Ether Products*, Midland, MI.

Foda, Y.H., Mahmoud, R.H., Gamal, N.F. and Kerrolles, S.Y. (1987) Special bread for body weight control. *Annals Agr. Sci.*, **32(1)**, 397–407.

Ganz, A.J. (1974) How cellulose gum reacts with protein. *Food Eng.*, June, 67–69.

Glicksman, M., Frost, J.R., Silverman, J.E. and Hegedus, E. (1985) US Patent 4,503,083.

Greminger Jr, G.K. and Krumel, K.L. (1980) Alkyl and hydroxyalkylcellulose (Chapter 3). In: *Handbook of Water-soluble Gums and Resins*, R.L. Davidson, ed., McGraw-Hill, New York.

Hansen, P.M.T., Hildalgo, J. and Gould, I.A. (1971) Reclamation of whey protein with carboxymethylcellulose. *J. Dairy Sci.*, **54(6)**, 830–834.

Heuser, E. (1944) *The Chemistry of Cellulose*, John Wiley, New York, pp. 379–391.

Higgins, T.E. and Madsen, D.P.D. (1986) US Patent 4,596,727.

Hood, H.P. (1981) Whipped sour cream in aerosol can. *Food Eng.*, **53(8)**, 62.

Huber, C.S. and Rowley, D.M. (1988) US Patent 4,737,374.

Huber, C.S., Rowley, D.M. and Griffiths, J.W. (1989) US Patent 4,826,656.

Ingram, P. and Jerrard, H.G. (1962) *Nature*, **196**, 57.

Isogai, A. and Atalla, R.H. (1991) Amorphous cellulose stable in aqueous media: regeneration from SO_2 – amine solvent system. *J. Poly. Sci.: Part A*, **29**, 113–119.

Jackman, K.R. (1979) US Patent 4,163,807.

Keeney, P.G. (1982) Development of frozen emulsions. *Food Tech.*, November, 65–70.

Keller, J. (1984) *Sodium Carboxymethylcellulose*, Special Report, NY State Agricultural Experimental Station, No. 53, 9–19.

Kester, J.J. and Fennema, O. (1989) An edible film of lipids and cellulose ethers barrier properties to moisture vapor transmission and structural evaluation. *J. Food Sci.*, **54(6)**, 1383–1389.

Kloow, G. (1985) Viscosity characteristics of high viscosity grade carboxymethyl cellulose (Chapter 32). In: *Cellulose and Its Derivatives: Chemistry, Biochemistry, and Applications*, J.F. Kennedy, G.O. Phillips, D.J. Wedlock and P.A. Williams, eds, Halsted Press, New York.

Klug, E.D. (1966) US Patent 3,278,521.

Klug, E.D. (1967) US Patent 3,357,971.

Klug, E.D. (1971) *J. Poly. Sci.: Part C*, **36**, 491–508.

Klug, E.D. and Tinsley, J.S. (1950) US Patent 2,517,577.

Krassig, D.H. (1985) Structure of cellulose and its relation to properties of cellulose fibers (Chapter 1). In: *Cellulose and Its Derivatives: Chemistry, Biochemistry and Applications*, J.F. Kennedy, G.O. Phillips, D.J. Wedlock and P.A. Williams, eds, Halsted Press, New York.

Luzietti, D. and Coacci, S. (1990) European Patent 354,356.

McCormick, C.L. and Callais, P.A. (1987) Derivatives of cellulose in LiCl and N,N-dimethylacetamide solutions. *Polymer*, **28**, 2317–2322.

Marmo, D. and Rocco, F.L. (1982) US Patent 4,311,720.

Michie, R.I.C. and Neale, S.M. (1964) *J. Poly. Sci.: Part A*, **2**, 2063–2083.

Midkiff, D.G., Twyman, N.D., Rippl, G.G. and Wahlquist, J.D. (1990) European Patent 0353334 Al.

Moore, L.J. and Shoemaker, C.F. (1981) Sensory textural properties of stabilized ice cream. *J. Food Sci.*, **46(2)**, 399–402, 409.

Ott, E. (1946) *High Polymers*. Vol. 5, *Cellulose and Cellulose Derivatives*, Interscience Publishers, New York.

da Ponte, D.J.B., Roozen, J.P. and Pilnik, W. (1987) Effects of Iota carrageenan, carboxymethylcellulose and xanthan gum on the stability of formulated minced fish products. *Int. J. Food Sci. Tech.*, **22(2)**, 123–133.

Sanderson, G.R. (1981) Polysaccarides in foods. *Food Tech.*, July, 50–57.

Seaborne, J. and Ebgerg, D.C. (1989) US Patent 4,820,533.

Shenkenberg, D.R., Chang, J.C. and Edmondson, L.F. (1971) Develops milk orange juice. *Food Eng.*, April, 97–98, 101.

Sirett, R.R., Eskritt, J.D. and Derlatka, E.J. (1981) US Patent 4,264,638.

SRI International (1989) *Chemical Economics Handbook*, Section 581.5000A (Cellulose Ethers), Menlo Park, CA.

Suderman, D.R., Wiker, J. and Cunninghan, F.E. (1981) Factors affecting adhesion of coating to poultry skin. *J. Food Sci.*, **46(4)**, 1010–1011.

Taguchi, A. and Ohmiya, T. (1985) US Patent 2,517,577.

Valbonesi, F. and Cochin, A. (1981) French Patent 2,478,955.

Vincent, A. and Harrison, S. (1987) Stabilizing dressings and sauces. *Food Trade Review*, October, 527–528, 531.

Ylimaki, G., Hawrysh, Z.J., Hardin, R.T. and Thomson, A.B.R. (1988) Application of response surface methodology to the development of rice flour yeast breads: objective measurements. *J. Food Sci.*, **53(6)**, 1800–1805.

4 Exudate gums

A.P. IMESON

4.1 Introduction

The exudate gums are amongst the oldest and most traditional thickening and stabilising agents used in food. Despite competition from other materials, several of these natural exudates continue to be used in large quantities. Indeed, in food, gum arabic is used more than any other polysaccharide except starch and its derivatives.

Many trees and shrubs yield gummy liquids which hydrate in water. These dry in the sun and air to form hard, glassy lumps. The shapes are characteristic for each material. Gum arabic is obtained as white to pale amber or brown striated nodules or tears. Gum karaya may form large, convoluted lumps, pale grey to dark brown in colour. Gum tragacanth is obtained from many different gum yielding species of *Astragalus* shrub. The exudates range from very white, long, narrow, curved ribbons to dark brown amorphous lumps for some flake grades.

The physiological basis for gum formation is unclear. Production is low in good climatic conditions but high temperatures and limited moisture increase gum formation. In addition, yields are improved by deliberately making incisions or stripping bark from the tree.

Over the last 20 years, major disruption to supplies of the food-approved exudate gums has been caused by political instability in Iran and Sudan, government intervention in India and recent severe droughts in the sub-Sahara regions. Limited availability and very high prices in the 1970s provoked a major loss of confidence by users and this was followed by a partial or complete switch to alternative materials. Concerted action in all the producing and exporting countries has established buffer stocks, introduced tapping controls with tree conservation and replacement, increased payments to collectors and funded research on productivity. These steps have stabilised supplies and prices and should avoid fluctuations in product availability and costs.

The particular properties of the different exudate gums have been identified and utilised in food applications over many years. They are used for emulsifying, thickening and stabilising a broad range of products. Arabic, tragacanth and karaya gums have a long, safe history of use in foods and recent toxicological data for each product has been assessed and approved.

Recent collaborative work between end-users, suppliers and researchers is generating much new data on the structure and function of these well-established gums. This is leading to a better understanding of gum selection and processing in order to obtain optimum product performance.

The individual gums, gum arabic, gum tragacanth and gum karaya, are discussed in order of their commercial importance to the food industry.

4.2 Gum arabic

4.2.1 Introduction

Gum arabic is the natural gum exuded by various species of *Acacia*. The main source of commercial gum arabic is *Acacia senegal* L. Willd., also called *Acacia verek*. Gum from *A. seyal*, known as gum talha, is not approved for food use in the US and Europe and it is mainly used in non-food products. Minor quantities of gum are obtained from *A. laeta* and other *Acacia* species.

The trees grow mainly in the sub-Sahara or Sahel zone of Africa but also in Australia, India and America. As well as providing the gum exudate, the trees play an important environmental role in stabilising the soil against erosion and reducing desert encroachment (Awouda, 1990). The main producing and exporting countries in the 'gum belt' include Senegal, Mali, Mauritania, Niger, Chad and Sudan.

Gum arabic has been accorded the highest possible status for a food additive of 'ADI not specified' following assessments of toxicological evidence by JECFA. This classification was also given by the US and the EC regulatory committees (Anderson and Eastwood, 1989). Recently, JECFA have reviewed the specification for this gum. They have proposed that the definition should be changed to 'the dried gummy exudate from *Acacia senegal* and closely related species' and for the specification to include permitted ranges for the specific rotation and for the Kjeldahl nitrogen value (Anderson *et al.*, 1991). These proposals are being studied at present.

Sudan dominates the gum trade. The gum is often called 'hashab', after the local name for the trees, or 'Kordofan' from the name of the main production area in the Sudan. The Gum Arabic Company (GAC) of Khartoum was established in 1969 with a monopoly for marketing this commodity in the Sudan. Until the climatic changes in the early 1970s, Sudan supplied about 75% of the annual demand for about 45 000 tonnes of gum arabic (Robbins, 1987). Failure by the GAC to warn customers of the pronounced shortage in 1973–1974 caused a major loss of confidence by manufacturers. Some reformulated with modified starches and dextrins and the demand for gum arabic fell.

The GAC introduced tapping and planting schemes, stabilised prices and set up buffer stocks to improve confidence in this gum. The consequence was a shift towards Sudanese supplies and from 1977–1984 it provided around 90% of the total world supply (Anderson and Eastwood, 1989). A severe drought in 1984 interrupted trade and led to a temporary quadrupling of prices. Supplies subsequently stabilised. However, since 1991, cold weather, foliage attack by locusts and changes in export duties on arabic have severely reduced exudation and gum supplies coming to market. Buffer stocks are now exhausted and only 8500 tonnes are expected in the 1992–1993 season to meet an annual demand for 28 000 tonnes. Importers expect to meet most demands but some qualities will be limited.

With normal supplies, estimates for future demand predict a small drop of about 1% per annum in total consumption in Europe (GIRA, 1990).

4.2.2 Manufacture

4.2.2.1 Collection and grading. Gum is exuded from *A. senegal* or 'hashab' trees in the form of large (5 cm diameter) striated nodules or tears. Mature trees, 4.5–6 m high and 5–25 years old, are tapped by making incisions in the branches and stripping away bark to accelerate exudation. The gum dries into rough spheres which are manually collected and taken to central markets. Collection takes place at intervals during the dry season from November to May and two main harvests are taken in December and April (Thevenet, 1988). In general, the higher the average temperature, the greater the production of gum. Nevertheless, the yield from each tree rarely exceeds 300 g. Gum production from wild stands of *Acacia* trees is gradually being replaced by the cultivation of uniform stands of monocultures of *A. senegal* in the Sudan (Awouda, 1990). Development programmes in the Kordofan and Darfur provinces are particularly advanced to ensure the continued supply of large quantities of pure gum from *A. senegal*. Cultivation is not widely practised in other producing areas so that gum arabic from other African countries, principally Nigeria with smaller amounts from Mali, Senegal, Mauritania, Niger, Burkina Faso, Chad, Tanzania and Kenya, may be variable in quality due to the mixed species of *Acacia* trees found in the collection areas.

Local collections are delivered to central markets for grading. At the main distribution centre of El Obeid in the Sudan the gum is sorted by hand into two main grades of 'hand-picked-selected' (HPS) and 'cleaned'. Other grades may be offered as shown in Table 4.1. Different grading systems are operated in the other exporting countries, such as Nigeria, and, although the Sudanese production is more tightly controlled, supplies

Table 4.1 Gum arabic:equivalent commercial grades.

Sudanese grade	Alternative terms	Nigerian grade	Powdered gum solution properties[†]	
			Clarity	Colour
—	Superior selected*	—	Clear	Very pale yellow
Hand-picked-selected (HPS)	Selected sorts	—	Clear	Pale yellow
Cleaned and sifted	Cleaned, sifted sorts	No. 1	Clear	Pale–dark yellow
Cleaned	Cleaned amber sorts or cleaned Kordofan	No. 1	Slight haze	Pale–dark yellow
Siftings	—	No. 2	Cloudy	Yellow–amber
Dust	—	No. 3	Opaque	Dark amber–brown

* Processed powder only.
† Spray-dried and roller-dried gum solutions are slightly hazy to cloudy and colourless to pale brown depending on gum arabic quality.

of good quality Nigerian No.1 grades from reputable dealers can match Sudanese material.

Variations in consignment quality and a lack of local cleaning facilities have been addressed by importers in the US, Europe and Japan who purchase material on the basis of approval of pre-delivery samples and then process to ensure rigorous product specifications are met.

4.2.2.2 Processing. Historically, gum arabic was simply sold in its natural state with little or no processing. Some simple treatments, such as granulating or grinding uncleaned material or sieving whole gum to remove sand and fine gum, were undertaken. Much raw gum continues to be sold but this is now pre-cleaned to remove bark, sand and fines. Material sold in kibbled, granulated or powder form is also usually pre-cleaned to improve quality. Local cleaning facilities in the exporting countries are generally inadequate and importers in the US and EC routinely process material to ensure materials meet specifications (Robbins, 1987). After cleaning, the bark and foreign matter is below 0.5% in food-grade powdered *Acacia*.

Additional processing by spray-drying and roller-drying commenced in the mid 1970s to give products with no insoluble matter and faster hydration properties. Both processes commence with kibbled gum arabic which is sieved several times, decanted after holding in a sedimentation tank, and centrifuged before drying (Williams, 1990). The concentrated solution is heated to pasteurise the gum and to reduce viscosity and facilitate handling. Holding times are minimised to reduce protein denaturation and loss of emulsification properties (see section 4.2.4.3).

Spray-dried gum is made by spraying the atomised solution into heated chambers and removing the particles through cyclones. Temperatures rarely exceed 80–90°C. The particle size and agglomeration of the spray-dried gum is controlled to optimise wetting, dispersing and hydration properties.

Roller-dried gum is produced by applying the solution to steam-heated rollers and removing the dried material with a knife. Roller-dried gum disperses easily and hydrates very rapidly because the material is in the form of relatively large flakes which have a large surface area. However, the high processing temperatures may denature the proteinaceous fractions in the gum and affect functional behaviour.

Both spray-dried and roller-dried products give slightly turbid or opalescent solutions indicating some change in properties as a result of heat processing. In addition, ^{13}C-nmr spectroscopy has given unequivocal evidence that occasional batches of processed gum contain different *Acacia* gums, such as *Combretum nigricans* or *A. seyal* (gum talha) which are processed with *A. senegal* (Anderson *et al.*, 1991).

4.2.2.3 Hygiene. Compared to the other exudate gums and other polysaccharides in general, gum arabic carries very low levels of microflora. Typically, total bacterial counts are around 1000 per gram and pathogens are absent (Blake *et al.*, 1988). Spray-dried gum arabic has lower counts due to the high temperatures reached during its manufacture and typical values are around 400 per gram.

For the vast majority of food applications, the low levels of microorganisms, coupled with pasteurisation or high temperature treatments during food processing, render sterilisation unnecessary. Where a reduction in viable bacteria is required, sterilising processes are readily available. Formerly, ethylene oxide gas treatment was used but this is no longer permitted in foods. Propylene oxide is used in the US but the gas is less efficaceous than ethylene oxide.

Heat treatments of stock solutions, for example during the manufacture of spray-dried or roller-dried gums, reduce the microflora. However, prolonged heating can lead to autohydrolysis of the naturally acidic solution and cause precipitation of the arabinogalactan–protein complex (Anderson and McDougall, 1987). As discussed in section 4.1.4.3, this highly functional component is responsible for promoting emulsification and stabilising a range of food products (Randall *et al.*, 1989).

Irradiation eradicates all viable coliforms, moulds and all other bacteria on raw, kibbled and spray-dried gum arabic using doses of 1, 5 and 8 kGy, respectively (Blake *et al.*, 1988). Irradiation is known to cause electrolytic scission of the glycosidic bonds in polysaccharides. However, at a dose of 10 kGy no significant changes in the molecular weight of the gum are detected and, although viscosity is reduced, there is no measurable effect on emulsion stabilisation. At present, irradiation offers the most effective process for sterilising raw gum arabic.

Table 4.2 Analytical data for gum arabic (*Acacia senegal*) (Anderson *et al.*, 1990) and comparison with other *Acacia* gums (Anderson, 1977).

	Acacia senegal		Acacia seyal	Acacia laeta	Acacia compylacantha	Acacia drepanolobium
	Average ± SD*	Test article†				
Ash (%)	3.8 ± 0.4	—	2.87	3.30	2.92	2.52
Nitrogen (%)	0.34 ± 0.03	0.31	0.14	0.65	0.37	1.11
Methoxyl (%)	0.24 ± 0.06	0.26	0.94	0.35	0.29	0.43
Specific rotation (degrees)	−30 ± 1.3	−30	+51	−42	−12	+78
Intrinsic viscosity (ml/g)	17 ± 2	17	12.1	20.7	15.8	17.8
Equivalent weight	1030 ± 70	1020	1470	1250	1900	1980
Uronic acid (%)	17 ± 2	17	12	14	9	9
Sugar composition after hydrolysis (%)						
4-O-methyl glucuronic acid	1.5 ± 0.5	1.5	5.5	3.5	2	2.5
Glucuronic acid	16 ± 5	15.5	6.5	10.5	7	6.5
Galactose	45 ± 5	45	38	44	54	38
Arabinose	24 ± 3	24	46	29	29	52
Rhamnose	13 ± 2	14	4	13	8	1

* Average of 35 samples ± standard deviation (SD).
† Test article used in the toxicological evaluation of gum arabic by JECFA.

4.2.3 Chemical composition

Gum from *A. senegal* is a slightly acidic complex polysaccharide obtained as a mixed calcium, magnesium and potassium salt (Anderson *et al.*, 1990). It has a molecular weight of around 580 000 daltons. Hydrolysis yields carbohydrate fractions of galactose, arabinose, glucuronic acid and rhamnose. The analysis of *Acacia* gums shows slight variation between samples of the same species and large differences between material from different species. These variations include nitrogen content, amino acid composition, uronic acid content and molecular weight. Detailed characterisation of gums from different *Acacia* species by Anderson and his group of co-workers is shown in Table 4.2. The different gums may be readily identified from the chemical composition and specific rotation values. The main product for food applications, gum arabic from *A. senegal*, is remarkably consistent in quality. Material collected over an 80 year period in the Sudan and from other sources including Nigeria, Mauritania, Kenya, Chad, Uganda, Niger gives similar analytical values. There is no evidence of compositional differences or changes in optical rotation arising from physiological adaptation of trees which survived the droughts in the periods 1973–1974 and 1984–1985. Any significant deviation from the data for *A. senegal* in Table 4.2 is a consequence of adulteration (Anderson *et al.*, 1990).

Hydrophobic affinity chromatography has identified three principal fractions in gum arabic: a low molecular weight arabinogalactan (AG), a very high molecular weight arabinogalactan–protein complex (AGP) and a low molecular weight glycoprotein (GI) (Williams *et al.*, 1990). These components represent approximately 88%, 10% and 1% of the molecule, respectively. The polypeptide is unevenly distributed between these fractions: they contain about 20%, 50% and 30%, respectively. Enzyme degradation of the polypeptide fraction of the AGP complex indicates that the protein is located at the exterior of the AGP unit (Randall *et al.*, 1988).

The overall conformation of the gum arabic molecule is described in terms of a 'wattle blossom' model (Connelly *et al.*, 1987) in which about 5 bulky AG blocks, each of around 200 000 daltons, are arranged along the GI polypeptide chain which may contain up to 1600 amino acid residues.

4.2.4 Functional properties

4.2.4.1 Viscosity and rheology.   The unusual solution properties of gum arabic are a consequence of the highly branched and compact structure. Other polysaccharides of similar molecular weight adopt extended conformations in solution and develop high viscosities. In contrast, solutions below about 10% gum arabic give low viscosities and Newtonian rheology

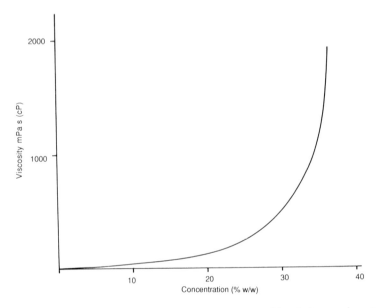

Figure 4.1 Viscosity as a function of concentration for gum arabic solutions measured with a Carrimed Controlled Stress Rheometer at $100\,s^{-1}$, 20°C (Williams *et al.*, 1990).

(Williams *et al.*, 1990). Above about 30% gum arabic the hydrated molecules effectively overlap and steric interactions result in much higher solution viscosities (Figure 4.1) and increasing pseudoplastic behaviour. Solutions of over 50% gum can be prepared and processed. This phenomenon allows the incorporation of high levels of gum in confectionery, encapsulated flavours, cereal products and bakery glazes. High gum levels assist processing by reducing the added water to a minimum so that drying is quickly achieved.

4.2.4.2 Acid stability. Gum arabic is stable in acid solutions, and products such as citrus oil emulsions exhibit good shelf stability. The natural pH of gum from *A. senegal* is 3.9–4.9 (Anderson *et al.*, 1990) resulting from the glucuronic acid residues. The addition of acid or alkali produces changes in solution viscosity as the electrostatic charges on the macromolecule alter (Figure 4.2) (Williams *et al.*, 1990). At lower pH values the reduced ionisation results in a more compact polymer volume and a lower solution viscosity. As the pH is raised the increased dissociation of the carboxylate groups extends the molecule giving a maximum viscosity around pH 5.0–5.5. Above this value, additional alkali raises the ionic strength of the solution which, in turn, masks the repulsive electrostatic charges regenerating the compact conformation with a lower viscosity.

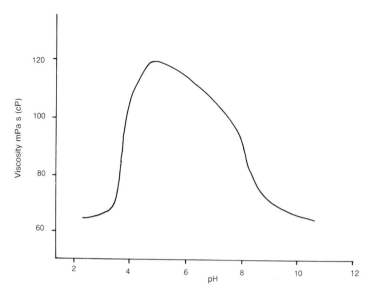

Figure 4.2 Viscosity as a function of pH for a 20% w/w gum arabic solution measured with a Carrimed Controlled Stress Rheometer at $100\,s^{-1}$, 20°C (Williams *et al.*, 1990).

4.2.4.3 Emulsification. The commercial exploitation of gum arabic for stabilising emulsions has been established over many years but the stereochemical basis for this application has only been elucidated in the last decade.

Gum from *A. senegal* contains a protein deficient, low molecular weight arabinogalactan (AG), a protein rich glycoprotein (GI) and a high molecular weight arabinogalactan–protein complex (AGP) (Williams *et al.*, 1990). Chromatography of a gum arabic solution and the aqueous phase of an orange oil emulsion stabilised with this gum has demonstrated that it is the high molecular weight AGP complex which is preferentially adsorbed on to the orange oil droplets. The AG complex shows no affinity for the oil phase and simply increases the viscosity of the aqueous phase. The AGP complex represents less than 10% of the gum arabic molecule and it contains about 20% protein (Randall *et al.*, 1988). Analysis showed that only about 1–2% of the gum is adsorbed at the oil–water interface and participates in the emulsification process. Consequently, concentrations of over 12% gum arabic are needed to give stable 20% w/w orange oil emulsions with a uniform small droplet size (Figure 4.3) (Williams *et al.*, 1990). Lower gum concentrations do not provide sufficient proteinaceous material to coat all the droplets fully so flocculation and coalescence occur. Higher oil levels or smaller droplet sizes need more gum arabic to give adequate stability.

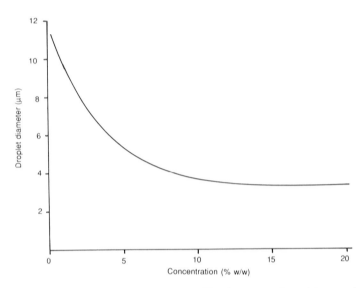

Figure 4.3 The average droplet diameter of a 20% w/w orange oil emulsion as a function of gum arabic concentration measured on dispersions diluted 1 : 10 000 with distilled water with a Coulter Counter with a 50 μm electrode (Williams *et al.*, 1990).

The AGP complex is believed to stabilise emulsions by absorption of the hydrophobic amino acids into the oil droplets. The large poly-saccharide blocks extend into the aqueous phase where steric repulsion between droplets prevents coalescence (Williams *et al.*, 1990).

A study with six different *Acacia* gums with nitrogen contents ranging from 0.09 to 7.5% demonstrated that emulsifying properties were cor-related with nitrogen levels (Dickinson *et al.*, 1988). Some variation in results was attributed to the distribution of the protein in the macro-molecule and its accessibility for adsorption into the oil–water interface. Gum arabic from *A. senegal* is the only material which has been toxicologically evaluated and approved for use in food. However, this gum only contains about 0.3% nitrogen and it is a relatively inefficient emulsion stabiliser. Nevertheless, other *Acacia* gums cannot be used in foods, despite their superior emulsification properties, unless tested and approved.

4.2.4.4 Heat stability.   Prolonged heating of gum arabic solutions leads to autohydrolysis of the naturally acidic solution (Anderson and McDougall, 1987). This gives a precipitate of the high molecular weight AGP and GI complexes and a concomitant decrease in the emulsification efficiency and solution viscosity (Randall *et al.*, 1989).

4.2.4.5 Compatibility and synergy.   Gum arabic is compatible with most gums and starches. Gelatin interacts at low pH values to give coacervates which have been utilised in the preparation of encapsulated oils.

Gum tragacanth and gum arabic mixtures give a synergistic viscosity decrease reaching a minimum at 80% gum tragacanth and 20% gum arabic. The mechanism for this is unclear but commercially the interaction has been used to give thin pourable emulsions with good shelf stability.

4.2.4.6 Sensory and nutritional properties.   The gum from *A. senegal* is generally odourless, colourless and tasteless except where the gum contains some associated bark and foreign matter. Spray-dried and roller-dried gums yield opalescent or turbid solutions as a result of heat processing.

In foods, about half of the gum arabic is incorporated as a minor additive, below 2% in the final product. The remainder of the gum is used in confectionery as a major ingredient at levels up to 60%. For these latter products an energy value of 14.7 ± 0.5 kJ/g (3.5 ± 0.1 kcal/g) should be used for calculating the energy contribution of the gum arabic component (Anderson and Eastwood, 1989). This value has been determined from feeding trials which confirm complete digestion of gum arabic by gut microflora, followed by absorption and utilisation of the gum during its passage through the body, when included at levels of up to 10% in the diet.

Previously, it has been considered that gum arabic, like most other polysaccharides, is poorly metabolised by the body and its calorific content is low. Indeed, this gum has been used in dietetic confectionery on the basis of its low digestibility and low energy value and tests have shown that this gum has a high fibre content of 94% and reduces serum cholesterol levels when ingested. However, the results of these feeding studies and opinions advanced by JECFA now support the higher figure for digestibility and energy value.

4.2.5 Applications

Gum arabic is readily soluble in cold and hot water and tends to form lumps when directly added to water. However, unlike viscous polymers, such as gum tragacanth and gum karaya, the low viscosity of gum arabic solutions allows these soft agglomerates to be dispersed and hydrated by continued agitation. To ensure rapid and complete hydration of this gum, it is still prudent to adopt the techniques used for mixing viscous polysaccharides in water:

- blend the gum arabic with other dry powdered ingredients, such as sugar

- slurry the gum in oil, glycerin or other non-aqueous fluid
- use a high-speed or high-shear stirrer
- disperse the powder with an eductor funnel and stir slowly
- use kibbled or whole gum for easy dispersion although complete hydration takes over 1 and 6 hours, respectively (Williams, 1990).

These methods are particularly useful for preparing stock solutions of more than 20% gum concentration where solution viscosities may be significant, such as for confectionery products. More details are given in section 4.3.5.

The food applications of gum arabic have been developed from its unequalled combination of properties. Emulsification, acid stability, low viscosity at high concentration, adhesive and binding properties and good mouthfeel characteristics have been used in five main food areas in descending order of importance:

- confectionery
- beverages and emulsions
- flavour encapsulation
- bakery products
- brewing

4.2.5.1 Confectionery. Gum arabic may be used alone or in combination with gelatin, agar or modified starch to produce various confections.

The most traditional products are the long-lasting, hard and chewy wine gums. Initially these were simply made from gum arabic solutions flavoured with wine (Best, 1990). Higher clarity could be achieved with gum arabic compared to other hydrocolloids and the resistance to melt-away, shape retention, bland taste and odour, pliable texture and low adhesion when chewed are all major benefits of this gum. Other properties of providing slow, controlled flavour release, protecting flavours from oxidation and controlling sugar crystallisation are also valuable.

Depending on the concentration of gum arabic, the sugar types and proportions and the residual moisture in the confectionery, textures ranging from soft lozenges and pastilles to hard gums can be made (Table 4.3) (Wolff and Mahnke, 1982).

Gum arabic shortages in the 1970s prompted various attempts to replace the gum with modified starches. These efforts were partially successful and blends of gum arabic and starch in proportions of 50:50 to 80:20 have been used. These products have a slightly different mouthfeel and texture but they continue to be used to reduce the dependency of the industry on gum arabic.

Chewy sweets are made using gelatin as the main texturising agent. Low levels of gum arabic (1.5–2.0%) are included to give additional

Table 4.3 Parameters for confectionery based on gum arabic (Wolff and Mahnke, 1982).

	Soft gums	Hard gums
Gum arabic (%)	30–35	40–55
Residual moisture (%)	15–17	10–13
Range of sucrose to glucose syrup ratio	65:50 to 50:50	70:30 to 65:35
Stoving time (hours)	24–36	48–60

adhesion, to reduce elasticity and to give extra fine sugar crystallisation with a smooth texture.

Gum arabic glazes are used on dragees, coated nuts and similar products.

For most confectionery applications, stock solutions are prepared by dissolving kibbled or whole gum into warm water up to 65°C. Higher temperatures cause cloudiness. The solution is filtered to remove foreign matter and then it is incorporated into a boiled solution of sugars. Flavours and colours are added before depositing and stoving for up to 4 days to obtain the desired moisture content (Reidal, 1986).

4.2.5.2 Beverages and emulsions. Gum arabic is widely used to stabilise emulsions containing citrus or fish oils and for emulsifying the flavour bases used for beverages.

The gum is completely hydrated in water. The oil is added in a controlled fashion and mixed with a high shear stirrer to form the initial emulsion. Stability is improved by reducing the oil droplets to an equal size of around 1 μm (Thevenet, 1988) and by using an oil weighting agent such as gum damar or gum elemi to increase the density of the oil closer to that of the beverage base mix.

A typical citrus oil emulsion for beverages will contain 6–8% flavour oil, 3–8% weighting agent and 15–20% gum arabic with the balance made up with water. High levels of gum arabic are used to ensure coverage of the oil droplet surface to prevent agglomeration. The beverage concentrate is completed by adding sugar syrups and citric or other acids. This concentrate is diluted about 5 times with carbonated water to give a drink with the required rheology, good flavour, mouthfeel and stability. The cloudiness of some grades of ground and spray-dried arabic is useful for providing a stable cloud in the drink and boosting the cloud from added fruit pulps and juices.

Fish oil emulsions are traditional health foods or dietary supplements stabilised with gum arabic and gum tragacanth. These gums give a synergistic viscosity decrease when used together so that the emulsion has very smooth, thin-pouring characteristics with very effective emulsification and long shelf stability.

4.2.5.3 Flavour encapsulation. Gum arabic has been recognised as an excellent encapsulating material for flavours because of its emulsification properties, low viscosity, bland flavour and for its protective action against flavour oxidation during processing and storage. Many dry foods, such as dessert mixes, soups and beverages, contain encapsulated flavours for flavour stability and longer shelf-life.

A typical formulation will contain 7% oil-based flavour and 28% gum arabic which gives 20% flavour in the dried material (Thevenet, 1988). Blends of 15% gum arabic and 25% maltodextrin with 10% flavour have also been used to give higher solids mixes which are processed faster and more cheaply and still give final products with 20% aroma. Higher concentrations of gum or maltodextrin give higher solution viscosities which cannot be readily spray-dried. It is important to ensure that the oil droplets are fully coated with gum prior to spray-drying otherwise volatile oils may be lost or oxidised. Microscopic examination of the dried encapsulated flavours shows particles ranging from 10 to 40 μm in diameter which contain numerous small (1 μm diameter) oil droplets.

4.2.5.4 Bakery products. Glossy coatings and the effective binding and sealing of baked goods and cereals often use gum arabic. The high solids and low viscosity of gum arabic are utilised by preparing concentrated stock solutions (30–50%) and spraying or brushing the coating onto the biscuit or pastry before baking. An attractive glossy coating forms as the gum solution evaporates. Dry roasted peanuts and almonds are amongst the applications for baked or roasted foods coated with gum arabic solutions.

In high sugar icings gum arabic binds water to help retain humidity and control moulding and rolling properties. When these glazes are applied warm to the baked goods, gum arabic maintains adhesion between the two surfaces.

4.2.5.5 Brewing. The charged uronic acid residues on this gum interact with the proteins in beers and lagers to stabilise the foam and assist lacing, the adhesion of the foam to the glass during drinking. Low levels of around 250 ppm of high quality powdered gum arabic are required to avoid cloudiness in the drink. The market tends to be dominated by propylene glycol alginate which is chemically synthesised to tight specifications. Potential exists for the use of pure, high quality gum arabic in this market as greater quantities of lagers and low alcohol drinks are developed.

In wine fining very low levels of gum arabic react with proteins to form flocs and sediments which can be removed by decanting or filtration.

4.2.6 Future developments

Gum arabic is used in foods in larger quantities than any other poly-saccharide. The combination of emulsification, good mouthfeel characteristics and low solution viscosities cannot be matched by any other gum and are only partially satisfied by some de-polymerised starches and dextrins. Advances with β-cyclodextrins may erode demand in flavour encapsulation but, for most significant food applications for gum arabic, satisfactory alternative materials have not been developed. Usage in traditional products may be affected if the markets for products such as hard gums and other confections change. If the end-user demands for a regular, consistent supply at current prices can be satisfied, the steady slow decline in the consumption of gum arabic in existing products should stop. Usage will increase in new applications in low-calorie beverages, emulsions and encapsulated products.

4.3 Gum tragacanth

4.3.1 Introduction

Gum tragacanth is the natural exudate gum obtained from small shrubs of the *Astragalus* species, comprising up to 2000 species, mainly located in south-west Asia. This plant is a small, bushy, perennial shrub with a large tap root which, together with the branches, is tapped for gum.

For regulatory purposes, gum tragacanth is defined as the dried exudation from the stems of *Astragalus gummifer* Labillardiere or other Asiatic species of *Astragalus*. The gum has been classified as generally recognised as safe (GRAS) in the US since 1961 and re-affirmed in 1974, following a long history of safe use in food and pharmaceutical products. A Joint WHO/FAO Expert Committee on Food Additives (JECFA) reviewed the available toxicological evidence and assigned it 'ADI not specified' in 1985. The European Community provisionally allowed gum tragacanth in food with an additive code E413 and a subsequent review of safety data, including animal and human feeding studies, by the Scientific Committee for Food, re-affirmed food additive status for gum tragacanth with 'ADI not specified' in 1988 (Anderson, 1989a).

The main producing areas for gum tragacanth are the arid and mountainous regions of Iran and Turkey. Commercial supplies are dominated by Iran which produces and exports 300 to 350 tonnes per year with an additional 80 to 120 tonnes per year from the Anatolia region in Turkey (Robbins, 1987). The gum has also been produced in Afghanistan and Syria but regular commercial supplies from these countries are unknown.

Historically, several thousand tonnes of tragacanth were used in food, pharmaceutical and technical applications. Following political upheaval in Iran in the late 1970s, greater central government influence led to very high prices from 1982 to 1985 (Robbins, 1987). At the same time there was strong competition from a new thickener, xanthan gum. Gum tragacanth usage fell dramatically. A loosening of state control since 1986 has seen prices fall but annual consumption has continued to drop to around 500 tonnes, with about 200–220 tonnes for food (Anderson, 1989a). Estimates for future usage predict a slow, continued fall of up to 6% per annum (GIRA, 1990) as existing traditional uses contract.

4.3.2 Manufacture

4.3.2.1 Collection and processing.   The stems, branches and tap root of the shrub are all tapped for the gum. The best quality is obtained from artificial incisions rather than natural exudations from the plant. Tapping or blazing is carried out in May or June with collection over a six week period in August and September for ribbon grades. Flake tragacanth is tapped and collected over a shorter period from August to November. A succession of tappings yields rapidly diminishing quantities of gum, so adequate supplies are maintained by visiting different sites.

The ideal climate consists of abundant rainfall prior to tapping and arid conditions during collection. Excessive rain and wind while the gum is being exuded results in a discoloured material with a lower solution viscosity.

Table 4.4 Equivalent commercial grades and viscosity of gum tragacanth.

	Iranian grade	Turkish grade	Approximate viscosity range	
			Redwood (s) (0.44%, 20°C)	Brookfield (mPa s) (1.0%, 25°C)
Ribbon	1		350–600	2200–3400
	2		250–400	1800–2500
	(3) Mixed		200–350	1400–2000
	4		120–170	1000–1600
	5	Fior	80–100	800–1000
Flake	(25)			
	26		70–85	600–800
	27	Bianca	65–75	400–700
	28		45–60	300–500
	(31)	Pianto		
	55		40–50	200–400
	(101)			
	(102)		30–35	20–30

Brackets indicate limited availability and commercial use.

After collection, the gum is sorted by hand into various grades of ribbon or flake. The Iranian grading system is more clearly defined than the Turkish and comprises about 14 different grades shown in Table 4.4. The most commonly used Iranian qualities are ribbons 1 and 4, mixed ribbon and flakes 27, 28 and 55. The best qualities are used where high viscosity, good solution colour and low microbiological limits are needed.

Processors in the USA and Europe purchase material following approval of pre-delivery samples. Quality control inspections of each incoming batch are necessary to ensure powder blends meet well-defined specifications for powder and solution colour and viscosity.

Most food applications for sauces, dressings, icings and confectionery will normally use mixed ribbon, various flake grades or the equivalent Turkish qualities, Fior or Bianca. Lower qualities are employed where solution colour is less important and where thermal processing, pH conditions and high solids are sufficient to ensure that no contamination is carried through to the finished food product.

4.3.2.2 Hygiene.   Limited cleaning to remove foreign matter may be undertaken in the exporting countries but no further processing is undertaken. Importers in the US and Western Europe, primarily the UK and Germany, ensure consistent quality standards are maintained. The best ribbon grades carry low total bacterial counts. These mainly comprise resistant spores from soil and airborne contamination. Previously, ethylene oxide gas treatment was used to reduce viable bacterial loads but this gas process is now no longer permitted in foods although it is still used for pharmaceutical products. In the US propylene oxide is permitted but its efficacy is limited.

4.3.3 Chemical composition

Gum tragacanth is a complex, heterogeneous, acidic proteoglycan of high molecular weight up to and in excess of 800 000 daltons. Hydrolysis yields arabinose, xylose, fucose, galactose, rhamnose and galacturonic acid, together with trace amounts of starch and cellulosic material (Anderson, 1989a). Under current definitions, gum tragacanth may be obtained from any Asiatic *Astragalus* species. Indeed, over 20 different species are used as gum yielders in Iran and Turkey, resulting in a wide variation in compositional analysis and functional properties of commercially available gum. Physical and chemical data for some commercial Iranian samples of gum tragacanth have been compiled by Anderson and Grant (1989). The data indicate that more viscous gum species contain high proportions of fucose, xylose, galacturonic acid and methoxyl groups and low proportions of arabinose and nitrogenous fractions. Low viscosity products contain more arabinose and galactose but galacturonic acid and methoxyl contents are lower.

Gum tragacanth consists of two fractions: a water-soluble, neutral arabinogalactan, tragacanthin, and an insoluble but water-swellable fraction, tragacanthic acid, also known as bassorin. The ratio of these components varies from species to species and partly explains the viscosity differences in commercial samples. Differences in soluble to insoluble fractions range from 90:10 in some samples of *A. echidnaeformis* to 50:50 for *A. shirineh* (Anderson and Grant, 1989). The nitrogenous fraction is also implicated in viscosity variation. The nitrogen content may range from 0.07% in soluble fractions of high viscosity material up to 1.87% in some insoluble fractions. This is equivalent to an average protein content ranging from 0.5 to 4.4%, which originates from peptide or protein sequences within the polysaccharide (Anderson and Grant, 1989). High hydroxyproline levels (up to 60%) in the amino acids may be involved in stabilising the structure of the arabinogalactan. This is indicated by the presence of small amounts of peptides rich in hydroxyproline in highly viscous gum tragacanth obtained from *A. microcephalus* and *A. gossypinus*. Peptide sequences in gum arabic are implicated in emulsion stabilisation and a similar functional role for the peptide-polysaccharide fraction in gum tragacanth is likely.

Carboxyl groups on the galacturonic acid residues in gum tragacanth are present as the calcium, magnesium and potassium salt forms.

4.3.4 Functional properties

4.3.4.1 Viscosity and rheology. At origin, gum tragacanth is simply graded by visual hand sorting. Suppliers carry out further necessary blending and grading to standardise products for viscosity and colour.

The gums hydrate in water to give viscous solutions at low concentrations and pastes at levels above 2 to 4%. A high quality ribbon tragacanth will give a 1% solution viscosity around 3500 mPa s although viscosities up to 4600 mPa s have been observed with some laboratory samples (Anderson, 1989a). Solutions are pseudoplastic with a reversible decrease in apparent viscosity as shear rate is raised. The high viscosity at low shear, coupled with charge repulsion effects from the galacturonic acid residues, suspend fine particles in solution and help stabilise oil in water emulsions.

4.3.4.2 Acid stability. Compared to most hydrocolloids, gum tragacanth is quite stable in acid solution. Figure 4.4 demonstrates the effect of acid on ribbon tragacanth. Flake grades show parallel stability but develop less viscosity. The stability is attributed to the resistance of the galactose backbone and its protection by side chains of arabinofuranose (Stauffer, 1980). Gum tragacanth solutions are acidic, normally between pH 5 to 6. Viscosity stability drops for solutions below pH 4 or above pH 6 but the

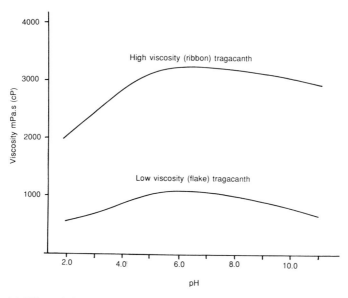

Figure 4.4 Effect of pH on viscosity of high viscosity (ribbon) and medium viscosity (flake) gum tragacanth (1.0%) solutions (Stauffer, 1980).

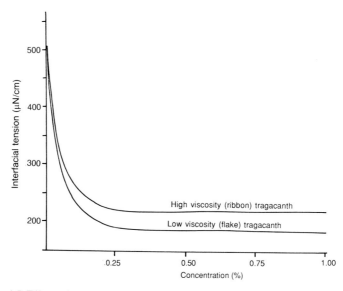

Figure 4.5 Effect of gum concentration on interfacial tension of oil–water emulsions (Stauffer, 1980).

acceptable stability of this gum at low pH led to its former widespread use in salad dressings, condiments, relish and other acidic products.

4.3.4.3 Emulsification. The surface tension of water is rapidly lowered by adding low concentrations of gum tragacanth (Stauffer, 1980). In oil and water mixtures the reduction in interfacial tension by tragacanth, shown in Figure 4.5, will facilitate emulsification and the viscous aqueous phase will help provide a stable emulsion. The data for surface tension and interfacial tension show the greatest effect with flake tragacanth. These lower viscosity grades have been found to contain higher nitrogen containing fractions compared to high viscosity samples (Anderson and Grant, 1989) and recent studies with gum arabic have identified the role of bound polypeptide or protein with surface activity and emulsifying properties (Dickinson *et al.*, 1988).

4.3.4.4 Compatibility and synergy. Most gums are compatible with gum tragacanth. Gum arabic produces an unusual viscosity reduction in gum tragacanth solutions. The mechanism is unclear but it is utilised commercially to produce thin pourable, smooth emulsions with fish and citrus oils which have a long shelf-life.

4.3.5 Applications

Solution preparation. In common with other cold water soluble hydrocolloids, powdered gum tragacanth tends to lump if directly introduced to water unless suitable precautions are used.

Hydration is a two-stage process. Satisfactory dispersion is a prerequisite so that water may be absorbed by the powder. Finer particles have a large surface area to volume ratio and fully hydrate faster than coarse material provided the dispersion is complete. Otherwise, the fine particles associate and effectively form large aggregates or lumps which hydrate very slowly. Coarse powders are selected where good dispersion is needed and short hydration times are unnecessary, such as for the preparation of stock solutions.

In dry powder mixes, or formulations which include sugar, maltodextrin, cook-up starch or seasonings, thoroughly blending the powdered gum with at least 5–10 times its weight of other ingredients will physically separate the individual gum particles. When added to water, the gum disperses well avoiding lump formation and ensuring rapid hydration and viscosity development.

Formulations containing oil, alcohol, glycerin or propylene glycol can be thickened readily by forming a slurry of gum in at least 5 times its weight of non-aqueous liquid and mixing this with water.

Stock solutions of the gum may be prepared by dosing the powder into the vortex of a rapidly stirred solution and using a high speed or high shear mixer to ensure good hydration and viscosity development. Bulk stock solutions may be prepared by using an in-line high shear mixer, or by evenly dosing the powder into water using a Venturi suction valve and slowly stirring the solution until the gum is hydrated.

Alternatively, kibbled gum tragacanth, which consists of 1–4 mm fragments of material, may be easily dispersed into solution without lumping. Slow stirring is continued for up to 24 hours at ambient temperatures until hydration is complete. Heating the solution to 50°C reduces the gum viscosity so solutions can be stirred more easily. At elevated temperatures solutions can attain peak viscosity within 2 hours. Caution must be exercised as excessive or prolonged heating can degrade the polymer and reduce viscosity.

Like other thickening and gelling agents, prolonged storage will lead to bacterial attack and polymer degradation. Combinations of heat treatment, refrigerated or frozen storage, low pH and preservatives, such as sorbic and benzoic acids and their salts or benzoic acid esters (p-hydroxybenzoates), are effective for maintaining solution properties throughout product preparation and shelf-life.

Food applications. As a result of its combination of functional properties, gum tragacanth has been employed in a wide range of foods. The thickening, acid and salt stability of gum tragacanth are similar to those of xanthan gum and many of the applications for these gums are similar. Other characteristics of gum tragacanth, such as the emulsification properties and creamy mouthfeel properties, are not matched by other gums and are used in specific applications. The food uses can be divided into five categories in descending order of importance:

- Confectionery and icings
- Dressings and sauces
- Oil and flavour emulsions
- Frozen desserts
- Bakery fillings

4.3.5.1 Dressings and sauces. Salad dressings, condiments, relishes and other low pH products are stabilised with gum tragacanth. Its primary emulsification properties are particularly effective in standard and low oil pourable salad dressings. The surface active properties of the gum emulsify and stabilise the oil droplets and provide a creamy mouthfeel. This is valuable for all relatively low viscosity pourable dressings, particularly for low- and no-oil products. The thickening and smooth flow behaviour gives long shelf stability by preventing separation of the oil phase while conferring even pouring and coating properties to the dressing. This range of properties may only be matched by using blends of propylene glycol alginate, gum arabic or cellulose derivatives for emulsification, together with xanthan and guar gums for thickening.

In condiments, such as mustard sauce, thickening, water binding, acid, salt and enzyme stability, together with the creamy mouthfeel and good flavour release properties, ensure gum tragacanth is retained as the preferred stabiliser in high value products.

In dressings and sauces the typical usage level is 0.4–0.8% of the weight of the aqueous phase depending on the oil content, the use of other thickeners and the consistency required. The gum is incorporated by dry blending with other ingredients or, more readily, by using a slurry in oil. In common with most gums, prolonged heating in acid conditions will degrade the polymer and lower viscosity. Processing conditions should be designed to minimise breakdown and maintain the valuable functional properties of the gum.

4.3.5.2 Confectionery and icings. Gum tragacanth is a very effective water-binding agent due to the high proportion of the insoluble but water-swellable fractions of bassorin. In aqueous products this component readily hydrates but in high solid systems the effective use of gum tragacanth is assisted by using fine mesh (75 μm or below) material.

Confectionery. Chewy sweets, such as lozenges, may be formulated with blends of:

- Gum tragacanth and gum arabic for a chewy texture
- Gum tragacanth and gelatin for a chewy and cohesive texture

These sweets are dry processed and stored in moisture proof packs or coated with chocolate for storage of hygroscopic tablets (Reidel, 1983). A peppermint lozenge may be produced by hydrating and kneading 2 kg gum tragacanth in 16 l of water. 500 g gelatin dissolved in an equal weight of warm water is added to this paste. Icing sugar (50 kg) is gradually worked into the paste until it is no longer sticky and finally 160 g of peppermint oil is added. The sweets are rolled and stamped into shapes before storing for 24 h at 35°C. The dried sweets are coated as required and packed. Lozenges flavoured with liquorice, menthol and herbal extracts are popular in Scandinavia and Northern Europe.

Gum tragacanth is also used in compressed confectionery, including fruit tablets and pastilles, in which it is used to bind the ingredients together during compression and to give a suitable consistency, mouthfeel and flavour release when chewed.

Icings. Icings are similar high sugar formulations in which gum tragacanth is the binding agent of choice. Fats are included in processed ready-to-roll icings to provide some pliability and reduce evaporative moisture losses when the product is used. The role of the gum is to:

- retain moisture during processing in order to facilitate machine rolling and extrusion
- maintain pliability throughout product shelf-life
- ensure cracks and breaks are avoided during hand or machine rolling
- give a smooth texture, good body and creamy taste during consumption.

4.3.5.3 Oil and flavour emulsions. Fish oil emulsions, sometimes flavoured with orange, are widely taken as dietary supplements for vitamins E and C throughout the Far East especially in China and Japan. Similar fish oil emulsions are available throughout Europe and the US and could be considered as some of the earliest 'functional foods' which are of increasing interest. The products have a rheology close to that of pourable dressings. Gum tragacanth and blends based on this gum are used for a combination of thickening, emulsifying and mouthfeel properties. Usage rates of 0.8 to 1.2% are used to give the long shelf-stability and flow properties needed.

Flavour oil emulsions, based on citrus fruits, are stabilised by gum tragacanth and its blends using the same technological justification as fish oil products.

4.3.5.4 Frozen desserts. In frozen confections, such as sorbets and ice lollies or ice pops, gum tragacanth can be used simply as a thickener in order to:

- control ice crystal growth
- reduce moisture migration and ice crystal development during storage
- prevent flavour and colour migration during storage and consumption.

In ice cream and sherbet, gum tragacanth helps to emulsify the fat in the mix as well as binding moisture, controlling ice crystal growth and providing body. Usage rates are around 0.2% for stabilising ice cream. Higher levels of 0.5% are used for sorbets and ice lollies and for providing stabilisation and some emulsification in ice cream.

4.3.5.5 Bakery fillings. The acid stability of gum tragacanth is employed in fruit fillings to give good clarity and gloss, together with a creamy texture.

4.3.6 Future developments

The broad and valuable combination of chemical and physical properties of gum tragacanth have resulted in its use in many foods. Competition from xanthan gum, which was only commercially available from the 1970s and which shares many of the characteristics of this gum, and guar gum, which is the most cost-effective thickener, have eroded much of the market share and application base held by gum tragacanth. The use of these gums and, most importantly, the synergistic viscosity of blends of these two materials, described fully in chapter 9, has particularly limited the use of gum tragacanth in dressings and sauces. Nevertheless, the properties of high viscosity, emulsification, acid and salt stability, shelf stability and clean, smooth and creamy mouthfeel properties maintain the well established position of gum tragacanth in traditional and premium quality products. In addition, niche applications, such as ready-to-roll icings for bakery products, cannot be satisfactorily matched by other gums or combinations. For these products, gum tragacanth will continue to be dominant.

The future continued use of this natural gum will depend on consistent qualities and stable prices being maintained. Consumer demands for higher quality foods, the growth of existing products and the development of new products, such as low calorie dressings, may slightly increase the demand from current levels.

4.4 Gum karaya

4.4.1 Introduction

Gum karaya, also known as sterculia gum, is defined as the dried exudate from trees of *Sterculia urens* (Roxburgh) and other species of *Sterculia*. The majority of commercial material is obtained from *S. urens* trees which grow wild in central and northern India. Other significant sources are from *S. setigera* in Senegal and Mali, with minor supplies from the Sudan and from *S. villosa* in Indian and Pakistan.

World production and usage is currently 3000–4000 tonnes per year. The major users of gum karaya are the US, France and the UK. Minor quantities are imported into Japan, Belgium, Germany and other European countries (Robbins, 1987). Out of the total gum karaya market it is estimated that 85–95% is used in pharmaceuticals as bulk laxatives, dental fixatives and colostomy sealing bags. World-wide only about 5% of gum karaya, amounting to less than 100 tonnes per annum, is used in foodstuffs (Anderson, 1989b).

Gum karaya has been classified as GRAS in the US since 1961 and this status was re-affirmed in 1974 (Anderson, 1989b). A subsequent review endorsed the use of the gum as a direct human food ingredient, subject to specified use level limits. In 1974 the European Community gave temporary approval for the use of gum karaya in foods and assigned it the code 416. Independent assessments of toxicological evidence sponsored by the International Natural Gums Association for Research (INGAR) led to an ADI value of 0–12.5 milligrams per kilogram body weight by the EC's Scientific Committee for Food, which has been upheld in more recent additive reviews. Despite these findings, gum karaya is not permitted throughout the EC: a ban remains in force in Germany. In contrast, JECFA granted gum karaya the status 'ADI not specified' in 1988. Human feeding trials have confirmed that this gum is not degraded and it passes through the body unmodified, simply acting as a bulk laxative (Anderson, 1989b).

Formerly, Indian trade dominated world markets and governed gum quality, availability and prices. The local industry was organised initially through private merchants. In the 1980s, state nationalisation through the National Association for Export Development (NAFED) and, subsequently, the Tribal Marketing and Development Federation of India (TRIFED), 'canalised' all gum collection, marketing and sales and set export prices. In addition, a conservation policy was introduced to secure the long term future of the gum by instigating local tapping bans and initiating replanting schemes (Robbins, 1987).

The immediate effect was a decline in the production of good quality gum which, together with bad weather conditions and pressure to increase collectors wages, forced a doubling of karaya prices from 1981 to 1984. Gum supplies and prices have now stabilised. Recent privatisation should return supplies to their fomer status. However, exports from India have fallen from over 6500 tonnes in 1978 to under 1700 tonnes in 1990. Tonnages from Senegal and other exporting countries have increased rapidly to satisfy demand and now amount to 1000–1500 tonnes per year (Anderson, 1989b). Increasing production and competition from African suppliers will further erode traditional Indian dominance and should restrict price fluctuations in future.

Demand for future usage in foods is expected to remain firm at current levels (GIRA, 1990).

4.4.2 Manufacture

4.4.2.1 Collection and processing. Gum karaya is obtained by tapping or blazing mature, large (10 m high) bushy *Sterculia* trees. Exudation commences immediately with most gum being obtained within the first

Table 4.5 Gum karaya: commercial qualities.

Indian or African grade	Colour	Bark and foreign matter (BFM) (%)
Hand-picked-selected (HPS) ⎤	White to very light tan	0–0.5
Superior no. 1 ⎦	or grey	1.0–2.0
Superior no. 2	Very light tan	1.5–3.5
Superior no. 3 (fair, average quality, FAQ)	Tan	2.5–4.0
Siftings	Brown	5.0–7.0

day. Yields of between one and five kilos are obtained at each tapping and up to five visits may be made to a tree during its lifetime (Meer, 1980). The dried gum is collected as large, irregular tears from April to June before the monsoon season in India. The Senegalese crop is harvested in September to January and March to July. At village collection points, the gum is manually or mechanically cleaned to remove adhering bark, chopped or broken and sorted on the basis of colour and residual foreign matter.

The grading system used for commercial qualities of gum karaya is given in Table 4.5. The best grades of hand-picked-selected and superior no.1 are used where good solution colour and high viscosity or moisture binding are required.

Uses of the siftings grade are limited because of the high bark and foreign matter content (BFM): for compliance with US National Formulary/Food Chemical Codex the BFM must not exceed 3%.

US and European processors purchase material on the basis of pre-shipment samples. Selective blending using data from lot inspection and analysis is used to manufacture products to meet defined commercial standards and customer specifications.

4.2.2.2 Hygiene. The microbiological levels in this gum are typical for exudate and seed gums and the low pH conditions in sauces and dressings and pasteurisation or other heat treatments for foods ensure that this natural gum can be widely used.

4.4.3 Chemical composition

Gum karaya is a partially acetylated, complex branched polysaccharide with a very high molecular weight of around 16 million daltons (Le Cerf *et al.*, 1990). Hydrolysis gives glucuronic acid, galacturonic acid, galactose and rhamnose in varying proportions depending on the species, quality and age of the gum (Meer, 1980). The polysaccharide contains around 40% uronic acid residues and up to about 8% acetyl groups. The presence of these latter substituents prevents the native gum nodules fully dissolving in water but allows them to swell. Chemical deacetylation using dilute ammonia or sodium hydroxide solution is able to modify the gum characteristics. The equivalent weight drops from 520 for the native gum to 460 for the deacetylated product (Le Cerf *et al.*, 1990) and it changes from a water-swellable to a water-soluble material.

Indian karaya from *S. urens* is characterised by a strong odour of acetic acid and has an acid value of 19.5 to 32.0. In contrast, African karaya from *S. setigera* has less odour and has a lower acid value of 12.5 to 23.0.

The structure of gum karaya is not fully characterised but it is believed to contain a central chain of galactose, rhamnose and galacturonic acid residues with side chains of glucuronic acid. The exudate occurs in the calcium and magnesium salt forms.

Low levels of amino acids have been detected in gum karaya (Anderson, 1989b), equivalent to about 1% proteinaceous material, lower than the levels found in other exudate gums.

4.4.4 Functional properties

4.4.4.1 Viscosity and rheology. Gum karaya absorbs water quickly to form a viscous colloidal dispersion at low concentrations. Native, acetylated gum karaya assumes a compact and branched conformation as the deformable particles swell in water (Le Cerf *et al.*, 1990). At concentrations of 2.0 and 3.0% in water viscosities approach infinity at low shear stress values, indicating a yield stress for these dispersions of 60 and $100 \mu N/cm^2$ respectively (Mills and Kokini, 1984). Hence, gum solutions will suspend particulates and give soft, spreadable gels with a jam-like consistency. In comparison, some thickeners, such as guar gum, do not form a network and flow under all shear stresses, whereas xanthan gum solutions have recognised suspension properties and have an elastic modulus (G') around $200 \mu N/cm^2$, about 300 times greater than guar at shear rates of 0.1 radians/s (Kelco, 1988).

Viscosity is affected by particle size and shear history. Fine particles absorb moisture quickly to give a smooth solution whereas coarse particles hydrate more slowly and give a grainy dispersion. The hydrated swollen particles are not stable to mechanical shear. Prolonged stirring gives smooth solutions with a reduced viscosity.

Deacetylation of the native product gives a more expanded conformation to the molecule. In solution, this increases gum solubility and the polymer behaves as a random coil generating higher viscosities (Le Cerf *et al.*, 1990). These solutions are cohesive and stringy or ropy. Indeed, solutions of African gum karaya, which has a lower acetyl content, are typified by their pronounced degree of ropiness.

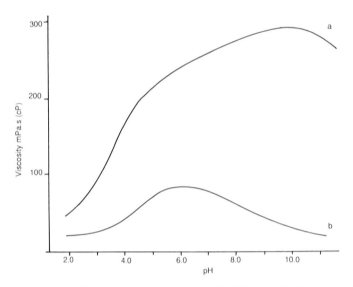

Figure 4.6 Effect of pH on gum karaya dispersions (0.5%) with pH adjustment after (a) or before (b) gum hydration (Meer, 1980).

4.4.4.2 Temperature stability. Pastes are produced at concentrations above 3 to 4%. Heating changes the polymer conformation and increases the solubility. This is accompanied by an irreversible viscosity loss so that smooth solutions up to 15% concentration may be made.

4.4.4.3 pH stability. The pH of an Indian gum karaya solution is about 4.4 to 4.7. As the pH is increased deacetylation occurs *in situ*. This is accompanied by a rise in viscosity and an increase in the degree of ropiness (Figure 4.6). African gum karaya naturally has a lower acid value which results in a higher solution pH of 4.7–5.2 and solutions have the characteristic ropy rheology and high viscosity of a deacetylated gum.

4.4.4.4 Adhesion. High concentrations of 20 to 50% gum karaya in water give heavy pastes with good adhesive properties. These are used in dental adhesives and colostomy bag sealing rings to develop and maintain bond strength, even when diluted.

4.4.4.5 Sensory properties. The characteristic odour and flavour of gum karaya is stronger for Indian rather than African material and is attributed to the acetyl groups which hydrolyse with time. These properties, coupled with the colour which ranges from light tan or grey to brown, restrict the use of this gum to applications where these characteristics are useful, such as in brown sauces or savoury bakery glazes, or where use levels are low and sensory properties are not affected.

4.4.5 Applications

Like gum tragacanth, powdered gum karaya rapidly absorbs water and swells. Mixtures may tend to lump unless appropriate techniques are used to prepare the gum karaya solution. Alternative methods include:

- Blending the gum karaya with other dry powdered ingredients such as sugar or maltodextrin.
- Slurrying the gum in oil, alcohol or glycerin.
- Hydrating the powder with a high shear or high speed mixer.
- Using a Venturi eductor funnel to disperse the powder in water and stirring slowly.
- Selecting a coarser particle size or kibbled (pea-size) gum to disperse easily and hydrate more slowly.

Combinations of techniques are frequently used to ensure effective hydration. More details on these methods are given earlier for gum tragacanth in section 4.2.5.

Gum karaya was originally exploited as a cost-effective alternative to gum tragacanth. The properties of these two cold water soluble thickeners differ in major aspects, primarily odour, taste, colour, rheology and resistance to mechanical shear, so successful opportunities for replacing gum tragacanth were limited. Specific applications have been developed for gum karaya. In many cases, subsequent reformulation with guar, xanthan or other gums has occurred. Current uses for gum karaya are found in the following product areas:

- Sauces and dressings
- Dairy products
- Frozen desserts
- Bakery products
- Meat products

4.4.5.1 Sauces and dressings.   Sauces and dressings are typical products for utilising the benefits of gum karaya, namely the high viscosity at low concentration, suspension properties and acid stability. The characteristic odour and flavour of this gum are acceptable in these acid products. Nevertheless, competition from guar and xanthan gums, alone or in combination, and, to a lesser extent, gum tragacanth, limit the products where gum karaya is the stabiliser of choice. Levels of about 0.6–1.0% are used to give the required opacity, colour, suspension and smooth consistency. The recipe and process are designed to accommodate the low temperature and shear stability of this gum.

4.4.5.2 Dairy products.   Gum karaya has been used as a binder in cheese spreads where it is used to reduce whey exudation, provide consistency and body and improve spreading characteristics. The gum has foam stabilisation properties which could be used in whipped cream and other aerated dairy products.

4.4.5.3 Frozen desserts. In frozen desserts, including sorbet, sherbet and ice lollies, low levels of 0.2 to 0.5% gum karaya are used to control moisture migration and bleeding, reduce suck-out of colour and flavour and control ice crystal size. In ice cream, gum karaya is not widely used but it can be used as a partial replacement for locust bean gum (Robbins, 1987). This latter material is used to control and limit meltdown, a property which would be provided by gum karaya of African origin which gives cohesive (ropy) solutions. The usage of gum karaya is limited by cost competitiveness and flavour against other gums in these applications.

4.4.5.4 Bakery applications. In common with other thickeners, gum karaya can be used in bakery formulations to improve tolerance to variations in water addition and mixing time. The moisture retention properties are used to reduce the effects of staling and to extend shelf-life in long life (over 7 days) baked goods.

Traditional recipes for coatings and glazes on baked products are thickened with karaya. This gum glaze adheres well to the pastry surface and gives additional colour after baking.

4.4.5.5 Meat products. In sausages and comminuted meat products gum karaya is included at low levels:

- to improve adhesion between meat particles
- to bind water during preparation and storage
- to impart an improved body and smooth texture.

Recent developments for diet or low-calorie burgers have included gum karaya for its physical and sensory properties and to provide soluble fibre.

4.4.6 Future developments

The properties of gum karaya include general attributes of thickening, suspension and water binding together with specific properties of acid stability, adhesion, odour, flavour and colour. This combination is difficult to match with other gums alone or in combination. Worldwide, the use of gum karaya in foods is low and it is unlikely to grow: its functional properties preclude its testing in most new product developments. However, the recent evaluations by JECFA and the EC Scientific Committee for Food underline the safety of this gum.

Future demand may increase if foods with high fibre/bulk laxative properties — a major pharmaceutical application — are required. In the absence of this development gum karaya usage will continue at or below current levels.

References

Anderson, D.M.W. (1977) Water-soluble plant gum exudates — Part 1: gum arabic. *Proc. Biochem.*, **12**(10), 24–25, 29.

Anderson, D.M.W. (1989a) Evidence for the safety of gum tragacanth (Asiatic *Astragalus* spp.) and modern criteria for the evaluation of food additives. *Food Additives and Contaminants*, **6**(1), 1–12.

Anderson, D.M.W. (1989b) Evidence for the safety of gum karaya (*Sterculia* spp.) as a food additive. *Food Additives and Contaminants*, **6**(2), 189–199.

Anderson, D.M.W. and McDougall, F.J. (1987) Degradative studies of gum arabic (*Acacia senegal* (L.) Willd.) with special reference to the fate of the amino acids present. *Food Additives and Contaminents*, **4**(3), 247–255.

Anderson, D.M.W. and Eastwood, M.A. (1989) The safety of gum arabic as a food additive and its energy value as an ingredient: a brief review. *J. Human Nutr. and Diet.*, **2**, 137–144.

Anderson, D.M.W. and Grant, D.A.D. (1989) Gum exudates from four *Astragalus* species. *Food Hydrocolloids*, **3**(3), 217–223.

Anderson, D.M.W., Brown Douglas, D.M., Morrison, N.A. and Weiping, W. (1990) Specifications for gum arabic (*Acacia senegal*): analytical data for samples collected between 1904 and 1989. *Food Additives and Contaminants*, **7**(3), 303–321.

Anderson, D.M.W., Millar, J.R.A. and Weiping, W. (1991) Gum arabic (*Acacia senegal*): unambiguous identification by ^{13}C-NMR spectroscopy as an adjunct to the revised JECFA specification, and the application of ^{13}C-NMR spectra for regulatory/legislative purposes. *Food Additives and Contaminants*, **8**(4), 405–421.

Awouda, E.H.M. (1990) Indicators for present and future supply of gum arabic. In: *Gums and Stabilisers for the Food Industry 5*, G.O. Phillips, D.J. Wedlock and P.A. Williams, eds, IRL Press at the Oxford University Press, Oxford, pp. 45–54.

Best, E.T. (1990) Gums and jellies. In: *Sugar Confectionery Manufacture*, E.B. Jackson, ed, Blackie, Glasgow, pp. 190–217.

Blake, S.M., Deeble, D.J., Phillips, G.O. and Du Plessey, A. (1988) The effect of sterilising doses of γ-irradiation on the molecular weight and emulsifying properties of gum arabic. *Food Hydrocolloids*, **2**(5), 407–415.

Connolly, S., Fenyo, T.-C. and Vandevelde, M.-C. (1987) Heterogeneity and homogeneity of an arabinogalactan–protein–*Acacia senegal* gum. *Food Hydrocolloids*, **1**(5/6), 477–480.

Dickinson, E., Murray, B.S., Stainsby, G. and Anderson, D.M.W. (1988) Surface activity and emulsifying behaviour of some *Acacia* gums. *Food Hydrocolloids*, **2**(6), 477–490.

GIRA (1990) *Hydrocolloids in western Europe, winners and losers in the 90s*, GIRACT SARL, Geneva, 670pp.

Kelco (1988) *Xanthan gum*, 3rd edn, Kelco, Div. of Merck & Co., Inc., San Diego, 23pp.

Le Cerf, D., Irinei, F. and Muller, G. (1990) Solution properties of gum exudates from *Sterculia urens* (karaya gum), *Carbohydr. Polym.*, **13**(4), 375–386.

Meer, W. (1980) Gum karaya. In: *Handbook of Water Soluble Gums and Resins*, Davidson, R.L., ed, McGraw-Hill, New York, chapter 10, 10.1–10.14.

Mills, P.L. and Kokini, J.L. (1984) Comparison of steady shear and dynamic viscoelastic properties of guar and karaya gums. *J. Food Sci.*, **49**(1), 1–4, 9.

Randall, R.C., Phillips, G.O. and Williams, P.A. (1988) The role of the proteinaceous component on the emulsifying properties of gum arabic. *Food Hydrocolloids*, **2**(2), 131–140.

Randall, R.C., Phillips, G.O. and Williams, P.A. (1989) Effect of heat on the emulsifying properties of gum arabic. In: *Food Colloids*, R.D. Bee, P.J. Richmond and J. Mingins, eds, Royal Society of Chemistry, Cambridge, pp. 386–390.

Reidel, H. (1983) The use of gums in confectionery. *Confect. Prod.*, **49**(12), 612–613.

Reidal, H. (1986) Confections based on gum arabic. *Confect. Prod.*, **52**(7), 433–434, 437.

Robbins, S.R.J. (1987) *A Review of Recent Trends in Selected Markets for Water-Soluble Gums*, Overseas Development Natural Resources Institute, Bulletin No. 2, 108pp.

Stauffer, K.R. (1980) Gum tragacanth. In: *Handbook of Water Soluble Gums and Resins*, R.L. Davidson, ed, McGraw-Hill, New York, chapter 11, 11.1–11.31.

Thevenet, F. (1988) Acacia gums, stabilisers for flavor encapsulation. In: *Favour Encapsulation*, S.J. Risch and G.A. Reineccius, eds, American Chemical Society, Washington DC, chapter 5, 37–44.

Williams, G.R. (1990) The processing of gum arabic to give improved functional properties. In: *Gums and Stabilisers for the Food Industry 5*, G.O. Phillips, D.J. Wedlock and P.A. Williams, eds, IRL Press at the Oxford University Press, Oxford, pp. 37–40.

Williams, P.A., Phillips, G.O. and Randall, R.C. (1990) Structure-function relationships of gum arabic. In: *Gums and Stabilisers for the Food Industry 5*, G.O. Phillips, D.J. Wedlock and P.A. Williams, eds, IRL Press at the Oxford University Press, Oxford, pp. 25–36.

Wolff, M.M. and Manhke, C. (1982) Confiserie la gomme arabique. *Rev. Fabr. ABCD*, **57**(6), 23–27.

5 Gelatin

J. POPPE

5.1 Introduction

For more than 2000 years connective tissues and products extracted from them have been used in the home and in the food industry for their gelling properties, and also as technical products in the form of adhesives. Availability and product consistency improved when industrial-scale gelatin manufacture appeared at the end of the nineteenth century. More recently, an improved knowledge of amino acids and proteins in general, and of collagen and gelatin in particular, together with the introduction of modern production techniques, has made it possible to produce gelatin which is bacteriologically safe and in accordance with international standards and rigid specifications.

The main areas in which gelatin is used, are:

- Food industry (confectionery, meat products, dairy products, etc.)
- Pharmaceuticals (capsules, etc.)
- Photography
- Technical applications.

In the food industry, gelatin is one of the hydrocolloids or water-soluble polymers that can be used as a gelling, thickening or stabilising agent. It differs from other hydrocolloids because most of them are polysaccharides, such as carrageenan and pectins, whereas gelatin is a totally digestible protein, containing all the essential amino acids except tryptophan.

At present, world production of gelatin is estimated to be 140 000–160 000 tonnes per year and there is an average annual increase in use of gelatin in the food area of about 3%, mainly in confectionery and low-calorie spreads.

5.2 Gelatin: definition

Gelatin is a high molecular weight polypeptide derived from collagen, the primary protein component of animal connective tissues, which include bone, skin and tendon (Ramachandran, 1967). The name gelatin came into common use in about 1700 and is derived from the Latin *'gelatus'*, meaning firm or frozen. Although the term gelatin is sometimes used to refer to other gel formers, it is properly applied only to collagen-derived protein materials (Rose, 1987).

Gelatin is defined in the US Pharmacopeia (USP) (United States Pharmacopeia, 1990) as a product obtained by partial hydrolysis of collagen derived from the skin, white connective tissue and bones of animals. Apart from the USP, gelatin has been defined in similar terms in other official documents. The World Health Organization Report No. 48 B (FAO/WHO, 1973) recommends identification and purity standards for edible gelatin and the classification of gelatin as a food by the Food and Agriculture Organization (FAO)/WHO was subsequently endorsed by the European Community (EC).

5.3 Collagen

This animal protein is the major structural component of white connective tissue fibres, and is present in all tissues and organs. It constitutes almost 30% of the total protein in both vertebrates and invertebrates and, under a microscope, appears as white opaque fibres, surrounded by other proteins and mucopolysaccharides.

The amino acid composition of mammalian collagen is remarkably constant: 18 of the 20 amino acids generally found in proteins are always present. Collagen is characterised by a high content of glycine, proline and hydroxyproline, with the latter amino acid representing 13–15% of the collagen. Hydroxyproline is practically specific for collagen and is only otherwise found at about 2% in elastin. The details of the chemical structure of collagen have been described by many authors (Babian and Bowes, 1977; Johns, 1977; Veis, 1978; Ledward, 1986; Rose, 1987; Stainsby, 1990).

The basic element in the configuration of collagen is tropocollagen which consists of three chains, each left-handed (three amino acids per turn and an identity period of 8.6 Å), intertwined like the strands of a cable and held together by hydrogen bonds (Karlson, 1965). These three polypeptide chains form a slight, right-handed superhelix. Tropocollagen has a molecular weight of 360 000, is 3000 Å long and has a diameter of 14 Å. Different combinations of the 3 polypeptide chains, giving rise to different types of tropocollagen molecules (type I, II, III, IV), have been described by Ashgar and Hendrickson (1982).

Tropocollagen molecules are chemically linked to form fibrils. Fibres arise when fibrils line up alongside each other, each one shifted by one quarter of its length along its neighbouring molecule. These fibres are stabilised by intermolecular cross-linking between a lysine or hydroxyproline residue and a lysine (or hydroxylysine) derived aldehyde, forming a labile aldimine bond which becomes stable as the collagen ages (Babian and Bowes, 1977).

5.4 Collagen–gelatin transformation

The conversion of collagen to gelatin is the essential transformation that occurs in gelatin manufacture. This transformation of highly organised fibres of collagen, which are insoluble in water, to a more depolymerised system called gelatin, which is soluble in water, has already been described in great detail by Veis (1964). The complexity of the collagen structure and the variety of chemical and enzyme treatments that can be applied in the manufacture of gelatin, explain the existence of a great variety of gelatin types.

Although modern production and quality control methods are used, commercial gelatin-making still relies on empirical experience. It involves hydrolysis catalysed by acid or alkali for gelatins that gel in water. Enzyme hydrolysis is used for so-called hydrolysed gelatin. The simplest way to transform collagen into gelatin is to denature soluble collagen. Thermal denaturation can take place by heating the collagen in neutral or slightly acidic conditions to about 40°C. At that point, the fibres and fibrils of collagen dissociate into tropocollagen units by the loss of the hydrogen bonds and hydrophobic bonds which help to stabilise the collagen helix (Johns and Courts, 1977). The next step in the hydrolysis of collagen consists in breaking the intramolecular bonds between the three chains of the helix.

This hydrolysis can achieve three results:

- the formation of three independent alpha chains
- the formation of a beta chain (two alpha chains linked by one or more covalent bonds) and an independent alpha chain
- the formation of a gamma chain (three chains linked by covalent bonds).

The main difference between the alpha, beta and gamma forms of gelatin is molecular weight. For the alpha form, molecular weight varies from 80 000 to 125 000 and for the beta form, from 160 000 to 250 000. The gamma form has a molecular weight of 240 000 to 375 000.

5.5 Gelatin manufacture

Gelatin is produced from biological materials, the nature of which varies greatly. As described earlier, the source for gelatin is collagen.

Commercially, almost the only sources of practical importance are hides, bones and pigskin. Industrial production of gelatin began in the US in 1850. At that time the main raw materials were hides and bones but production increased greatly in 1930 when pigskin was introduced as a raw material. Manufacture in Europe began some 25–30 years later but Europe has now become the most important area for the production of gelatin.

With regard to the raw materials a distinction should be made between bone (ossein) gelatin, hide gelatin and pigskin gelatin, as the raw material used can determine the process adopted and the products formed.

Commercial gelatins can be divided into two groups: gelatin type A obtained by acid pretreatment and gelatin type B obtained by basic pretreatment. Some raw materials, such as bones, can be pretreated by both procedures but others, such as pigskin, are mainly processed by one method in particular.

5.5.1 Pretreatment

Industrial preparation of gelatin involves the controlled hydrolysis of the organised structure of collagen to obtain soluble gelatin. This can be done by an alkaline or an acid process.

Alkaline process. Demineralised bones (ossein) or hides are immersed in an alkali (usually lime or, in some cases, sodium hydroxide bath) for a number of weeks at ambient temperature. The programmed renewal of the bath liquor ensures the effective elimination of greases, fragile proteins and mucopolysaccharides and other minor organic compounds. Apart from this purification, an important purpose of the liming process is to destroy certain chemical cross-linkages still present in the collagen and, in doing so, render the collagen soluble in water. The alkaline process may last from 6 to 20 weeks.

Acid process (acidulation). In recent times the acid processing of pigskin and ossein has become more and more important in Europe. This procedure is particularly suitable for less cross-linked raw materials, such as the bones of young cattle and pigskin. The main reason for the success of this procedure is its short duration (10–48 h) in contrast to the lengthy alkaline process.

In this process the washed raw material is soaked in a dilute acid bath which contains a maximum of 5% of a mineral acid such as hydrochloric acid, sulphuric acid or phosphoric acid. The pH varies from 3.5 to 4.5 and the optimum temperature is 15°C. The acid treatment is stopped once the raw material is fully acidified or has reached maximum swelling. At the end of the process the excess acid is removed and the treated raw material is washed with cold water.

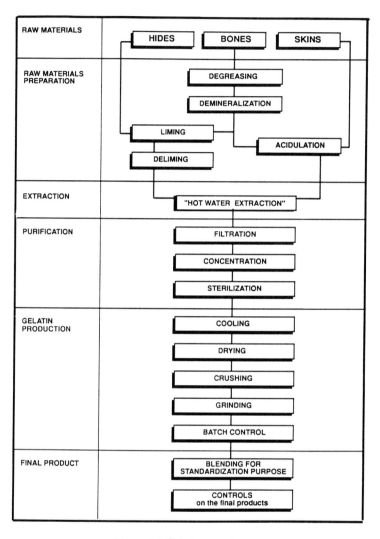

Figure 5.1 Gelatin manufacture.

5.5.2 Extraction and post-extraction processing

The extraction procedures for acid and alkali pretreated raw material
are very similar as can be seen in Figure 5.1. The collagen in the pretreated raw materials is digested in tanks using successive water extractions, the temperature of which is strictly controlled, varying from 55°C to
90°C. Different extractions are made in the same vessel. Each extraction
is carefully filtered, first through earth filters and then through cellulose

Table 5.1 Gelatin composition.

Protein	85–90%
Humidity	8–13%
Ash	0.5–2%

Table 5.2 Characteristics of gelatin type A and type B.

Type	A	B
Raw material	Pigskin, bone	Bone, hide
Pretreatment	Acid	Lime
Isoelectric point	7–9.4	4.5–5.3

sheets, in order to eliminate any impurities remaining in suspension, as well as the last traces of grease and albuminoid substances that have coagulated during extraction. The solutions are then concentrated in vacuum evaporators at sufficiently moderate temperatures to ensure that no degradation or loss of physical properties occurs. At the end of this operation, the concentration reached (25–45%) will depend on the nature of the raw materials and on the extraction conditions.

Once concentrated, the gelatin solution is subjected to a flash sterilisation at 140°C. It is then rapidly cooled so that it can be extruded in gel form. This is best carried out in an aseptic area where filtered and conditioned air is regularly controlled. The extruded gel passes through a continuous drier supplied with filtered air at a constant temperature. At the end of the drying process each successive extract is ground.

More details on gelatin manufacture are given by Hinterwaldner (1977) who describes not only the classic operating procedures of commercial gelatin manufacture but also some special treatments using hydrotropic agents, gamma rays and continuous processing.

5.5.3 Commercial gelatin

Commercial gelatin can be defined as a pure animal protein with the composition shown in Table 5.1. According to pretreatment, gelatin can be divided into gelatin type A and gelatin type B. The main differences between these two types are shown in Table 5.2.

For food applications the main differences are:

- Viscosity: for the same gel strength, type A gelatin is less viscous than type B.

Table 5.3 Amino acid composition of gelatin: amino acids obtained after hydrolysis of 100 g sample.

Residue	Weight (g)
Glycine	26–31
Alanine	8–11
Valine	2.6–3.4
Leucine	3–3.5
Isoleucine	1.4–2
Phenylalanine	2–3
Tryptophan	—
Serine	2.9–4.2
Threonine	2.2–2.4
Tyrosine	0.2–1
Proline	15–18
Hydroxyproline	13–15
Methionine	0.7–1
Cysteine	—
Cystine	Trace
Lysine	4–5
Arginine	8–9
Histidine	0.7–1
Aspartic acid	6–7
Glutamic acid	11–12
Hydroxylysine	0.8–1.2

- Isoelectric point: for blends with negatively charged hydrocolloids, such as carrageenans, gelatin type B should be used.

5.6 Chemical structure

5.6.1 Composition

All the amino acids that occur in proteins are present in gelatin with the probable exception of trypophan and cystine, although the latter is sometimes detected in trace amounts. Table 5.3 shows a typical analysis of the amino acids obtained after the hydrolysis of 100 g of gelatin. These percentages can vary and depend mainly on the raw material and, to a lesser extent, on the manufacturing process.

Amino acids are linked together in gelatin by peptide bonds. A typical sequence for gelatin is

Gly-X-Y

where X is mostly proline and Y is mostly hydroxyproline. This structure is shown in Figure 5.2.

Different peptide bonds result in a spatial configuration that is com-

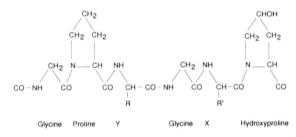

Figure 5.2 Chemical configuration of gelatin.

parable to the amino acid configuration in collagen. Indeed, it is very unlikely that any substantial chemical rearrangement would occur during the collagen–gelatin transformation. X-ray diffraction analysis has shown some fibril-like structure in gelatin. Nevertheless, this structure cannot be compared to the well ordered network of collagen (Bohonec, 1974).

5.6.2 Isoelectric point

In common with other proteins, gelatin can act either as an acid or as a base, depending on the pH. In an acidic solution the gelatin is positively charged and in an alkaline solution it is negatively charged. The intermediate point, where the net charge is zero, is known as the isoelectric pH or point.

Variation in the proportion of carboxylic acid amide groups is responsible for the differences in the isoelectric point of gelatin. In collagen, 35% of the acidic groups are in the amide form. Consequently, collagen is a basic protein with an isoelectric pH of 9.4 (Eastoe, 1967). During the preparation of gelatin, the acid or basic treatments hydrolyse the amide groups to a greater or lesser extent.

The isoelectric point of gelatin can vary between 9.4 (no modification of amide groups) and 4.8 (90–95% of free carboxylic acid groups) (Eastoe et al., 1961). Acid processed gelatins have high isoelectric points because the less severe processing conditions maintain the value near to that of collagen. Lime processed gelatins are subjected to a prolonged basic treatment and only a small proportion of the amide groups survive. The isoelectric point of these gelatins is acidic and remarkably constant between 4.8 and 5.2.

5.6.3 Molecular weight and molecular weight distribution

Molecular weight distributions can be determined by gel filtration chromatography, polyacrylamide gel electrophoresis and HPLC. An

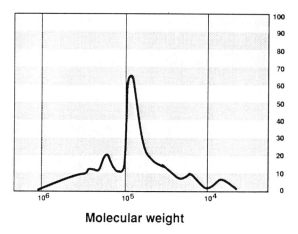

Molecular weight

Figure 5.3 Molecular weight distribution of gelatin (225 bloom limed hide).

example is given in Figure 5.3 for a high bloom, limed hide gelatin. High bloom gelatins normally contain a high proportion (30–50%) of molecules of similar size in the form of alpha and beta chains. Most gelatins also contain aggregates with a molecular weight up to 10 million and polypeptides with a molecular weight of less than 80 000 (Rose, 1987).

5.7 Functional properties

5.7.1 Gelation mechanism

Gelatin swells when placed in cold water, absorbing 5 to 10 times its own volume of water. When heated to temperatures above the melting point, the swollen gelatin dissolves and forms a gel when cooled. This sol–gel conversion is reversible and can be repeated. This characteristic is used advantageously in many food applications. Moreover, gelatin gels start to melt between 27 and 34°C and they tend to melt in the mouth. This is a desirable property in many foods.

The basic mechanism of gelatin gelation, which is now well accepted, is the random coil–helix reversion (Djabourov, 1989). The imino acid rich regions of the different polypeptide chains adopt a helical conformation on cooling and these helices are stabilised by hydrogen bonding which gives the three-dimensional gel. The gelation of gelatin can be considered

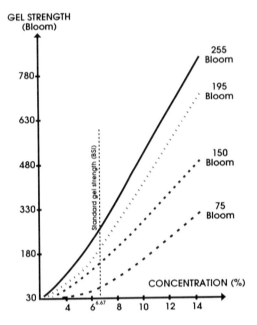

Figure 5.4 Gel strength of gelatin as a function of concentration.

to be a partial reformation of the collagen and these reformed parts act as the junction zones of the gel (Ledward, 1990).

5.7.2 Gel strength

The ability to form a gel is without doubt one of the most important properties of gelatin. The need to define and evaluate the characteristics of the gel has resulted in the concept of gel strength. The gel strength is, by definition, the weight in grams which it is necessary to apply to the surface of a gel, by means of a piston 12.7 mm in diameter, in order to produce a depression 4 mm deep. The jelly must be contained in a standard flask at a concentration of 6.67% and have been matured at 10°C for 16–18 h.

A standard method for the determination of the gel strength, based on the use of the bloom meter is described by the British Standards Institute (BS 757, 1975). Gel strength depends on the concentration of gelatin (Figure 5.4). Commercial products have gel strengths (bloom strength or bloom) between 50 and 300 bloom (grams) for a 6.67% gelatin concentration which increases with time as the gel 'matures', and varies inversely with temperature.

Hydrolysis, which can be provoked by numerous factors — acids,

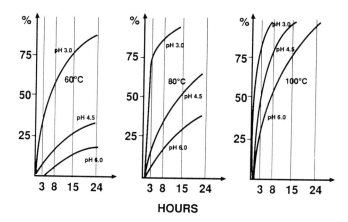

Figure 5.5 Loss of initial bloom gel strength (%) as a function of time, pH and temperature.

bases, temperature, enzymes, bacteria and irradiation — causes a progressive drop in the gelling properties of the gelatin. Variations in gel strength of a 210 bloom strength acid gelatin as a function of temperature, pH and time are shown in Figure 5.5 which gives the loss of bloom strength as a percentage of the initial bloom strength. The extent of the hydrolysis was followed by gel strength measurements made using the following procedure:

- (i) Aqueous solutions were prepared at 6.67% concentration by a 60 min cold water soak followed by melting down at 60°C for 10 min.
- (ii) The pH was adjusted with citric acid or sodium hydroxide to the appropriate value.
- (iii) Solutions were maintained at temperatures of 60°C, 80°C and 100°C in sealed flasks for 3 h, 8 h and 24 h. Finally, gel strength measurements were made according to the British Standard tests after 17 h at 10°C.

The loss of bloom strength is exponential, therefore the hydrolysis speed decreases as the hydrolysis progresses.

5.7.3 Melting point

In the BS 757 (1975) method, the melting point is the temperature at which a gelatin gel softens sufficiently to allow carbon tetrachloride drops to sink.

Melting point is affected by:

- the maturing time: Stainsby and Taylor (1958) indicated that the melting point of high grade gelatin is dependent upon conditions in the first hour of maturing
- the concentration: melting point increases with increasing concentration of the gelatin
- salts: sodium chloride depresses the melting point of gelatin gels.

The melting point of a 10% gelatin gel can vary from 27 to 32°C depending mainly on the bloom strength of the gelatin and the type of pretreatment of the raw materials.

5.7.4 Setting point

The setting point of a gelatin solution depends on its mechanical and thermal history. Mechanical action may delay setting and the temperature of setting is higher when the sol is cooled slowly compared to when it is chilled quickly (Stainsby, 1977a). The setting point of a 10% gelatin solution can vary from 24 to 29°C depending mainly on bloom strength and type of pretreatment used.

5.7.5 Viscosity

In general, viscosity of gelatin is measured from the flow time of a gelatin solution through calibrated viscosity pipettes. The most commonly used method in Europe, the BS 757 (1975) method, recommends the use of a glass capillary viscosimeter calibrated to BS 188. In this method a $6\frac{2}{3}\%$ solution of gelatin is prepared as for bloom testing (7.5 g of gelatin in 105 ml of water) and the flow time is measured at 60°C. The viscosity of this solution varies from 1.5 to 7.5 mPa s.

Stainsby (1977b) stated that the viscosity of a concentrated gelatin solution depends mainly on hydrodynamic interactions between gelatin molecules; the contribution from the solvent and from the individual gelatin molecules becomes less and less important as the concentration rises.

Viscosity also depends on temperature (above 40°C the viscosity decreases exponentially with rising temperature), on pH (viscosity is a minimum at the isoelectric point) and on concentration. Figure 5.6 shows the exponential variation of viscosity with concentration.

5.7.6 Turbidity

Gelatin turbidity is expressed in nephelometric turbidity units (NTU). The clarity of a gelatin solution depends mainly on the extraction and post-

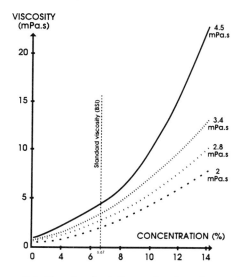

Figure 5.6 Viscosity of gelatin as a function of concentration.

extraction conditions. In general, the early extractions contain the highest quality of gelatin with a very good clarity. The later extracts, in contrast, can be turbid and more intensely coloured. The quality of the later extracts can be improved by clarification, bleaching and filtering. Gelatin solutions have a maximum turbidity at the isoelectric point.

5.7.7 Colour

The colour of gelatin depends on the raw materials, the processing methods and the number of the extraction. Colour is assessed by visual observation and comparison with a range of control gelatins. The results are expressed in Hellige units and can vary from 1.5 (pale yellow) to 14 (brown). Colorimeters measuring transmittance are commonly used for a numerical characterisation of the colour of gelatin solutions and gels.

5.7.8 Colloid protection

The protective colloidal action of gelatin results from its orientation at the interface between the two phases where it forms a monomolecular film round the colloidal particles. It also increases the viscosity of the aqueous phase and assists the formation and stability of suspensions and emulsions. This property is used in particular in the preparation of pro-cessed sauces, precooked dishes and desserts, and in acid milk desserts.

5.8 Uses of gelatin in the food industry

Gelatin is used in a great number of food applications. These can be divided into five groups:

1 Confectionery and jelly desserts
2 Dairy products
3 Meat products
4 Hydrolysed gelatin applications
5 Miscellaneous uses including sauces, dressings, wine fining, etc.

Gelatin is used because of its unique physical properties rather than for its nutritional value as a protein. The thermoreversibility of gelatin gels makes them unique. They have a melting point below 37°C which means they melt in the mouth and are easily dissolved.

The preparation of gelatin solutions presents few difficulties for concentrations below 10%. For levels up to 40–45%, three methods are used to prepare aqueous solutions of gelatin.

- Conventional method: first swell in cold water, then dissolve by heating in a water bath
- High-speed method: dissolve the gelatin in hot water using appropriate high-speed stirring
- Intermediate method: swell in cold water, then directly dissolve with other raw materials.

Particle size, gel strength, viscosity, concentration and dissolution time will determine the most appropriate method. The two most commonly used techniques are summarised in Figure 5.7.

The uses of gelatin can be classified according to function (Jones, 1977), i.e. gelatin can be considered as a:

Gelling agent: for jellied confectionery, aspic
Whipping agent: in aerated confectionery, aerated dairy desserts
Stabiliser: for ice cream, icings
Emulsifier: for salad dressings, whipped creams
Thickener: in flavouring syrups, canned soups
Adhesive: in confectionery (e.g. sticking together the different layers of 'liquorice allsorts')
Binder: in sugar pastes, liquorice
Fining agent: for wine, fruit juice

In this chapter the main uses of gelatin will be classified according to application area.

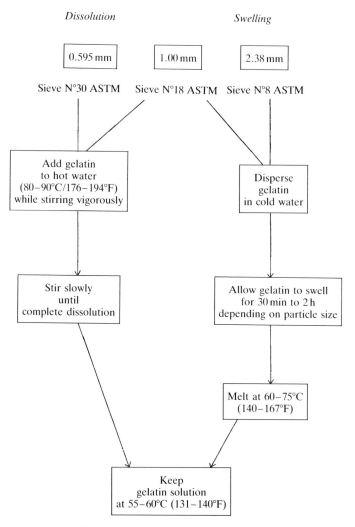

Figure 5.7 Dissolution of gelatin.

5.8.1 Confectionery

5.8.1.1 Dessert jelly. Table jellies are desserts made from sucrose and glucose and/or invert sugar syrup, gelatin, fruit acid(s), buffer, flavouring and colour which, after dissolving in hot water and cooling, set to form a clear jelly. The major consumers are North America and the UK. While many jellies are eaten in the home, institutions such as hospitals and schools are also major outlets for these products.

Two forms of presentation are available: jelly crystals and tablet jellies. Jelly crystals are a dry mix of the ingredients, while in tablet jellies the ingredients have been previously dissolved and set into a gel. In both cases the preparation of the final dessert involves the addition of hot water. Jelly crystals are more common in the USA than in the UK, where, traditionally, tablet jellies are the major product.

5.8.1.2 Jelly confectionery. The manufacture of confectionery jellies is similar to that of tablet jellies except that the confectionery is cast in depressions in moulding starch. Soft jellies may contain from 6 to 9% gelatin of various grades from 150 to 250 bloom. In general, type A gelatin is used in these applications because of its low viscosity. Soft jellies are ready to remove from the starch after one night and are then treated by various methods to prevent them from sticking together. The texture of the finished product is rather chewy. The best known example is the Gummi bear. A typical formula for 100 g of this type of product contains 7–8 kg gelatin 200 bloom type A dissolved in 14–16 kg water, 37 kg sucrose, 45 kg glucose syrup, acid colour and flavour. The solids content at depositing is around 78° Brix and the final solids content varies from 80 to 83° Brix.

An increasing number of these products are manufactured by continuous pressure-cooking manufacturing processes. In these cases, the gelatin solution is added to the sugars and glucose syrup before cooking and the cooking time is short enough to avoid gelatin hydrolysis.

Different textures and firmness can be obtained by:

- Varying the gelatin content; the more gelatin, the chewier the product
- Replacing one gelatin by another with a different gel strength; for the same firmness in the finished product high bloom gelatin gives less chewy products.

Gelatin can be combined with other stabilisers or gelling agents, to obtain particular textural characteristics, e.g.:

- gelatin with agar and pectin gives a short, brittle texture
- gelatin with modified starch gives a less elastic texture for wine gums, liquorice, etc.
- gelatin with gum arabic gives a hard, compact texture.

Gelatin is used as a texturiser in the manufacture of sugarless confectionery. The confectionery industry has been trying for years to replace sugar and glucose syrup in confectionery in order to produce candies that are non-cariogenic, suitable for diabetics and, if possible, reduced in calorie content. Today, polyols (sugar alcohols) are the only substitutes in the confectionery industry which provide the bulking effect necessary to give texture and body to the finished product and also provide a suitable sweetness. More and more, from a technological point of view, sugar and glucose syrup can be replaced by mixes of different polyols (lycasin–sorbitol, mannitol–sorbitol, xylitol–lycasin, etc.). Compared to classical formulae, it is necessary to add 10–15% more gelatin to the product, mainly due to the high hygroscopicity of most of the polyols used.

The influence of sugar and glucose syrup on the gelation of gelatin at the level of solids normally used in confectionery (78–85° Brix) has not been studied extensively. Oakenfull and Scott (1986) found that sucrose at 500 g/kg increased the melting temperature by 3.9°C. Marrs (1982) noted that the gelling properties of gelatin were modified by addition of glucose syrups and maltodextrins. Gelation was inhibited in the presence of low DE hydrolysates above a critical concentration. The type and dose of glucose syrup can influence considerably the texture of the finished product. This should be taken into account when developing a confectionery formula in which gelatin is the main gelling agent.

5.8.1.3 Aerated confectionery. Aerated confectionery can be defined as an aerated gelled product containing a mixture of carbohydrates: mainly sucrose and different types of glucose syrup, whipping and/or stabilising agents, flavour and colour. Some examples are moulded or extruded articles, such as marshmallows, soft foam articles and meringues, and confectionery which is cut into pieces by machine and individually wrapped, e.g. fruit chews, nougat and toffees.

Gelatin in aerated confectionery is used as:

- A foaming agent: the presence of gelatin decreases the surface tension of the liquid phase
- A stabiliser: the presence of gelatin gives the cell walls the necessary mechanical resistance to avoid deformation of the finished product
- A binder: gelatin is able to bind a large amount of water. It can help to give the finished product a longer shelf-life.

    Typical examples of aerated confectionery products containing gelatin are discussed in detail here.

Marshmallow. Marshmallow can be poured into a starch mould or extruded. This product generally contains between 2 and 5% of high gel strength gelatin. The texture of aerated articles can vary according to:

(i) the type and quantity of gelatin used:
- the rigidity of the product increases with gelatin content
- gelatins with high gel strength give short, creamy textures

(ii) the ratio of sugar to glucose syrup.

Products of the marshmallow-meringue type have a hard, relatively brittle texture. This is the result of a high sugar content, compared to glucose syrup, facilitating recrystallisation of the sugar. This phenomenon can be accentuated by drying out the products.

Fruit chews.   These confectionery products are characterised by:

- low aeration
- high total solids
- incorporation of fats

Gelatin is used in these products:

- for its foaming action
- to improve chewability
- to improve fat dispersion
- to control crystallisation of the sucrose

Toffees.   Toffees are manufactured from a mixture of milk, sugar, glucose syrup, fats and flavours, to which a frappé can be added after concentrating the sugars. A small quantity of gelatin is used to improve fat dispersion and stabilise the texture of soft toffees.

Nougat.   Nougat comprises a mixture of sugars, glucose syrup, in certain cases honey, invert sugar and fats, with the addition of a frappé after sugar concentration. Gelatin is a component of this frappé and is used to stabilise the texture of the finished product.

The main characteristics of gelatins used in gelled and aerated confectionery are summarised in Table 5.4.

5.8.2 Dairy products

Gelatin is used in the dairy industry as a texturising agent. Moreover, it has interesting foaming properties even in the presence of fat. Gelatin can be added to the milk before fermentation and its low melting point (under 37°C) gives the finished product a characteristic mouthfeel.

Gelatin is used in a number of milk products. Table 5.5 summarises the products and gives the characteristics and dose of gelatin to be used. The main examples include yogurts, fermented milks, milk gels, dessert creams, ice cream, low-calorie spreads and mousse.

Table 5.4 Characteristics of gelatin for confectionery.

Application	Gel strength (bloom)	Characteristics				Rate of use
		High viscosity	Average viscosity	Low viscosity	Others	
Jelly articles	175–250		X	X	Colour, brightness	6–9%
Hard gums	100–150			X		10–15%
Wine gums	100–150		X	X	Brightness	4–8%
Liquorices	75–125		X	X		3–8%
Marshmallows	200–250		X			2–5%
Deposited soft meringues	100–150		X	X		2–5%
Meringues	75–125		X			2–5%
Extruded articles	100–125	X	X			3–7%
Fruit chews	100–150		X		Foaming capacity	0.5–2.5%
Hydrolysed gelatin						
Caramels	100–150		X			0.2–1%
Nougats	100–150		X			0.2–1%
Dessert jelly	150–250		X		Transparency, colour	1.5–3%

Table 5.5 Characteristics of gelatin for dairy products.

Application	Gel strength (bloom)	High viscosity	Average viscosity	Low viscosity	Others	Rate of use
			Characteristics			
Long-life cream	150–250		X			0.2–0.5%
Whipped cream	150–250		X			0.2–0.6%
Low-calorie spread	225–250		X		Melting point	0.5–3%
Dry mix for flan, pudding	150–250		X			0.2–2%
Dry mix for mousse	150–250 Hydrolysed gelatin		X		Foaming	0.2–2%
Dry mix for bakery filling cream	150–250		X			0.2–3%
Flan, milk gel	150–250		X			0.2–2%
Foamed dessert, mousse	150–250 Hydrolysed gelatin		X		Foaming	0.2–3%
Dessert cream	150–250		X			0.2–1%
Ice creams and sorbets	100–250		X			0.2–1%
Yogurt-based products	100–150 Hydrolysed gelatin		X		Foaming	0.2–2%
Acid dessert	150–250		X			0.2–2%
Fresh curd-based products	150–250		X			0.2–2%

Yogurts. Gelatin is used to improve the texture of fermented yogurt, without affecting the characteristic taste of this product. The firmness of the finished product depends mainly on the amount of gelatin added. Gelatin is used mainly to prevent exudation of whey during handling and storage, and is particularly suitable for incorporation in fruit yogurts where exudation is largely inevitable unless a stabilising agent is added. However, it should be noted that the use of gelatin in yogurt products is limited in some countries by legislation.

Thermally-treated fermented milks. In certain countries, the tendency is to extend the shelf-life of certain types of yogurt and fermented milk by additional pasteurisation after the incubation process. This operation prevents prolonged action by the fermenting agents, but does destabilise the product texture leading to exudation. The addition of a mixture of gelatin and starch before pasteurisation produces a good texture and obviates any exudation effect. The texture of a product stabilised with gelatin alone is sensitive to changes in storage temperature. A combination with modified starch (0.4–0.6% gelatin with 1–1.5% modified starch) gives a highly satisfactory formulation in which the starch stabilises the viscosity between 5 and 20°C and avoids the risk of exudation.

Flavoured gelled milk desserts.   These dessert products have a semi-solid consistency and are prepared from flavoured sweetened milk. The stabilising agent employed must solubilise during thermal treatment of the milk, produce no increase in viscosity at high temperature and gel when the product is cooled in the pot. Gelatin can be used alone or in combination with other gelling agents, such as carrageenan.

Dessert creams.   Dessert creams have a relatively thick consistency and are made from flavoured, sweetened milk. Gelatin is used to achieve a smooth gel texture and prevent exudation on freezing or as a result of major temperature variations during storage.

Ice creams and water ices.   The presence of a stabilising agent is essential for the following reasons:

- to adjust mix viscosity
- to hold the emulsion in a stable state until the ice cream is consumed
- to facilite aeration and improve expansion
- to avoid deterioration during storage
- to prevent the formation of ice crystals during prolonged storage.

Used in association with other stabilising agents, gelatin provides the finished product with a remarkably slow melting rate and characteristic texture.

Low-calorie spreads.   The low-calorie spread market is expanding rapidly. The products concerned have reduced fat content and are either exclusively milk fat based, vegetable fat based or a combination of the two types of fat.

The preparation of a stable emulsion requires the presence of stabilisers and emulsifiers. The stabiliser must ensure good water binding and improve the structure, consistency and spreadability of the finished product. Good stabilisation can be obtained with high bloom gelatin, added at doses varying from 1 to 3%, mainly because the melting point of that type of gelatin (31–33°C) is very near to the melting point of the different fats present.

Other hydrocolloids (e.g. pectin, carrageenan and xanthan) may be used in association with gelatin in order to improve water binding.

Mousses.   The foaming action of gelatin is used in the preparation of a wide range of aerated products. Used in combination with other hydrocolloids, gelatin provides excellent emulsion stability, facilitating

Table 5.6 Characteristics of gelatin for meat products and miscellaneous applications.

Application	Gel strength (bloom)	High viscosity	Average viscosity	Low viscosity	Others	Rate of use
Meat industry jellies	150–250	X	X		Transparency, colour	3–15%
binders for meat emulsions	150–250	X	X			0.5–3%
hams, canned meat products	150–250	X	X		Transparency	1–2%
coatings	150–250		X		Transparency	5–20%
Fish products binders	150–250		X			0.5–3%
aspics	150–250		X		Transparency, colour	3–15%
Sauces, soups	150–250	X	X			0.5–2%
Wine clarification	75–125 Hydrolysed gelatin		X			

aeration and good mix stability, avoiding separation of the ingredients and maintaining foam stability before gelation.

5.8.3 Meat industry

Meat products hold an important position in the food industry in many countries. Gelatin is obtained from animal raw materials and can be considered a natural complement of meat-based proteins. Gelatin is frequently incorporated in the following:

- cooked pressed ham, cooked shoulder
- canned meat products
- meat emulsions (pork products, patés)
- jellies and aspics

The typical quality and dose of gelatin for each product is given in Table 5.6.

Cooked pressed ham, cooked shoulder. Gelatin is added to the ham during preparation, when the bone has been removed. Gelatin is also used to bond the meat where the rind and excess fat have been trimmed. The gelatin powder absorbs moisture from the meat and, during the cooking process, forms a film which seals the meat after cooling. The gelatin gels the liquid which exudes during cooking, and holds it in and around the ham. This added gelatin also stiffens the jelly obtained directly from the connective tissues during the cooking process to give an attractive presentation, and bind the ham together for improved slicing.

Canned meat products. Gelatin is used to gel the juices lost from meat products during cooking or pasteurisation. Gelatins with high gel strength (200–250 bloom) are used for this application at levels of 0.5–2%. As sterilisation temperatures are relatively high, gelatin gel strength and viscosity losses during processing should be taken into account as these essentially depend on thermal treatment duration.

Meat emulsions. As a result of their high fat and water content, these products present particular stabilisation problems, mainly from exudation of water or fat and texture irregularities after cooking. The use of a cutter partially eliminates these problems but gelatin is still used to achieve a better water binding, to stabilise the emulsion and to obtain a homogeneous batch texture.

Gelatin content varies considerably, depending on the presence of other binding agents, the amount of collagen present in the other ingredients, and local regulations.

Decorative jellies. Jellies are widely used for coating and decorating hams, pates, etc. Conventional jellies on a gelatin base or jellies prepared with type B gelatin and carrageenan are used for this application. The latter combination gives a clear jelly with:

- reduced setting time
- increased gel strength
- higher melting point (the addition of 10% carrageenan to the gelatin increases the melting point from 30 to 53°C).

5.8.4 Hydrolysed gelatin applications

Gelatin can be hydrolysed using enzymes to yield molecular weights down to 5000–10 000. In general hydrolysed gelatins are spray-dried, soluble in cold water and a 12.5% solution will not gel at 10°C.

Hydrolysed gelatin can be used in several ways:

(i) In confectionery as a foaming agent, either:

- in combination with a high bloom gelatin; it facilitates a short beating time and gives a very short texture to the marshmallow
- to stabilise fruit chews; the presence of hydrolysed gelatin

ensures minimum deformation while conditioning the finished product.

(ii) In dietetic food as a protein source: hydrolysed gelatin can be used in high protein dietary products but the absence of tryptophan is the limiting factor for gelatin in this type of application.

(iii) In granulation as a binder: this type of gelatin is used because it is cold soluble, has a low visosity and does not modify the physical, chemical or organoleptic properties of the powder treated by granulation.

(iv) in wine fining: used in a similar fashion to low bloom gelatin.

5.8.5 Miscellaneous applications

Gelatin is used at low levels in a lot of food applications: as a thickener in soups and sauces, as a stabiliser for salad dressings, a precipitant in wine fining and fruit juice clarification, and in microencapsulation of flavours, etc. For most of those applications, gelatin is combined with other proteins or hydrocolloids.

Emulsified sauces.   Several authors (Jones, 1977; Wood, 1977) quote the emulsifying power of proteins, especially of gelatin. Kragh (1958) studied the emulsifying and stabilising power of various gelatins used on their own in oil and water emulsions with different oils and fats. It can be assumed (Schut, 1976a) that the chains of the protein molecules are orientated at the oil/water interface, and that the configuration of the protein corresponds to a minimum free energy at which most polar groups interact with the polar groups of the aqueous phase, while the non-polar groups interact with the oil phase. A monomolecular layer of protein is thus formed around the oil droplets, stabilising the emulsion. Obviously, the mechanical properties of this monomolecular layer of gelatin and its resistance to external stresses will broadly determine the stability of the two-phase system.

According to Acton and Saffle (1971), the emulsifying power of gelatin is closely linked to its ability to generate a film around the dispersed phase and by increasing the viscosity of the oil in water emulsion (Krog and Lauridsen, 1976), the latter will be stabilised. This technique is used for emulsified sauces containing less than 80% of oil. In this case, gelatin increases the viscosity of the continuous aqueous phase and stabilises the sauce by binding excess water (Schut, 1976b).

Wine fining.   The operation of fining consists of incorporating a substance into the cloudy wine to flocculate and sediment the suspended particles. This substance is generally a protein such as gelatin, albumen or casein. The protein is precipitated by tannin or by the acidity of the wine, forming aggregates which readily sediment. The fining, provided it is done correctly, using pure products at low doses and with the minimum of aeration, does not modify the organoleptic qualities of the wine. Low bloom gelatin (75–125 bloom) and hydrolysed gelatin can be used for this purpose. Doses vary from 7 to 15 g per hectolitre.

Fruit juice clarification. The same principles are applied as for wine fining but the doses of gelatin are higher and vary from 20 to 120 g per hectolitre.

References

Acton, J.C. and Saffle, R.L. (1971) Stability of oil in water emulsions. *J. Food Sci.*, **36**, 1118–1120.

Ashgar, A. and Hendrickson, R.L. (1982) Chemical, biochemical, functional and nutritional characteristics of collagen in food systems. *Adv. Food Res.*, **28**, 233–251.

Babian, G. and Bowes, J.H. (1977) The structure and properties of collagen. In: *The Science and Technology of Gelatin*, A.G. Ward and A. Courts, eds, Academic Press, London, pp. 1–27.

Bohonec, J. (1974) An investigation of the supermolecular structure of dried gelatin layers. *Colloid and Polymer Sci.*, **252**, 333–334.

BS 757 (1975) *Methods for sampling and testing gelatin*, British Standards Institution.

Djabourov (1989) A review on gelatin gelatin: recent experiments and modern concepts. *Photogr. Gelatin Proc. 5th IAG Conf. (56 TEAA) 1988*, **1**, 133–146.

Eastoe, J.E. (1967) *Composition of Collagen and Allied Proteins: Treatise on Collagen*, Vol. 1, Academic Press, pp. 1–67.

Eastoe, J.E., Long, J.E. and Wilan, A.L.D. (1961) *Biochem. J.*, **78**, 51–56.

FAO/WHO (1973), Report No. 48 B, WHO/Food Add./70.40.

Hinterwaldner, R. (1977) Technology of gelatin manufacture. In: *The Science and Technology of Gelatin*, A.G. Ward and A. Courts, eds, Academic Press, London, pp. 315–361.

Johns, P. (1977) The structure and composition of collagen containing tissues. In: *The Science and Technology of Gelatin*, A.G. Ward and A. Courts, eds, Academic Press, London, pp. 32–66.

Johns, P. and Courts, A. (1977) Relationship between collagen and gelatin. In: *The Science and Technology of Gelatin*, A.G. Ward and A. Courts, eds, Academic Press, London, pp. 138–173.

Jones, N.R. (1977) Uses of gelatin in edible products. In: *The Science and Technology of Gelatin*, A.G. Ward and A. Courts, eds, Academic Press, London, pp. 366–392.

Karlson, P. (1965) *Introduction to Modern Biochemistry*, 2nd edn, Academic Press, New York, p. 50.

Kragh, A.M. (1958) The emulsion stabilizing properties of gelatine. In: *Recent Advances in Gelatine and Glue Research*, Pergamon Press, pp. 231–235.

Krog, N. and Lauridsen, J.B. (1976) *Food Emulsifiers and Their Associations with Water in Food Emulsions*, Stig Friberg, p. 117.

Ledward, D.A. (1986) Gelation of gelatin. In: *Functional Properties of Food Macromolecules*, J.R. Mitchell and D.A. Ledward, eds, Elsevier Applied Science, London, pp. 171–174.

Ledward, D.A. (1990) Functional properties of gelatin. In: *Gums and Stabilisers for the Food Industry 5*, G.O. Phillips, D.J. Weslock and P.A. Williams, eds, IRL. Press at Oxford University Press, Oxford, pp. 145–156.

Oakenfull, D. and Scott, A. (1986) Stabilisation of gelatin gels by sugar and polyols. *Food Hydrocolloids*, **1**(2), 163–175.

Ramachandran, G.N. (1967) *Treatise on Collagen*, Academic Press, New York.

Rose, P.I. (1987) Gelatin. In: *Encyclopedia of Polymer Science and Engineering*, 2nd edn, pp. 488–513.

Schut, J. (1976a) *Meat Emulsions in Food Emulsions*, Stig Friberg, p. 427.

Schut, J. (1976b) *Meat Emulsions in Food Emulsions*, Stig Friberg, p. 450.

Stainsby, G. (1990) Source and production of gelatin. In: *Gums and Stabilisers for the Food Industry 5*, G.O. Phillips, D.J. Weslock and P.A. Williams, eds, IRL. Press at Oxford University Press, Oxford, pp. 133–143.

Stainsby, G. (1977a) The gelatin gel and the sol-gel transformation. In: *The Science and Technology of Gelatin*. A.G. Ward and A. Courts, eds, Academic Press, London, pp. 179–206.

Stainsby, G. (1977b) The physical chemistry of gelatin in solution. In: *The Science and Technology of Gelatin*, A.G. Ward and A. Courts, eds, Academic Press, London, pp. 109–135.

Stainsby, G. and Taylor, J.T. (1958) *GGRA Research Report A20*.

United States Pharmacopoeia (1990), Official Monographs for USP XXII/NF XVII.

Veis, A. (1964) *Macromolecular Chemistry of Gelatin*, Academic Press, New York.

Veis, A. (1978) Collagen. In: *Encyclopedia of Food Science*, M.S. Peterson and A.H. Johnson, eds, Air Publishing Co., pp. 153–158.

Wood, P.D. (1977) Technical and pharmaceutical uses of gelatin. In: *The Science and Technology of Gelatin*, A.G. Ward and A. Courts, eds, Academic Press, London, p. 424.

6 Pectins

C.D. MAY

6.1 Occurrence and sources of pectin

The gelling power of pectin has been used in foodstuffs ever since the first fruit preserves were made. Pectin itself was first isolated and named by Braconnot (1825), who carried out some of the first systematic studies on the subject. Pectin continued to be of academic as well as practical interest, and a brief review of early work is given by Kertesz (1951). Pectic materials occur in most land plants, especially in soft tissues such as young shoots, leaves and, above all, fruits. In plants, they have an important role in the middle layer of the plant cell wall, helping to bind cells together, in association with cellulose, hemicelluloses and glycoproteins. Many of the more academic papers on pectic substances are concerned with this wide class of materials.

Despite this wide occurrence, only a few source materials have been used to produce commercial pectin as an additive for use in foods. One of the reasons for this is that many of the pectic materials present in nature do not have the necessary functional properties, especially the ability to gel acid sugar systems, which has been the main requirement in commercial pectins until recently. The broad class of pectic substances includes many with high proportions of neutral sugar units in the polymer chain, or highly substituted with acetyl or other groups, and these frequently prevent gelation.

Equally important is the fact that the pectin industry has developed as a by-product industry, using waste materials from the food industry, especially from the production of fruit juices and juice-based drinks. These wastes are the main source of large quantities of readily available pectic substances. The development of the pectin industry follows from the early practice of making an extract from waste materials such as apple peels and cores, and adding it to preserves, both domestically and in industry, to produce a good set. However, this was not always convenient for the growing preserves industry, as such trimmings were not available at the time of greatest demand, for example in the strawberry season, and thus the practice of drying the apple residue or pomace obtained from juice manufacture developed, particularly in Germany. An extract could then be prepared from this pomace when it was required.

Although initially the extract was prepared as required by the jam maker, it was soon realised that the extraction could be optimised by operating as a specialised business. The extract was concentrated, preserved with sulphur dioxide and sold in barrels. A successor to this operation, supplying 'liquid pectin' in bulk and in polythene barrels, still continues in the UK today.

It is not possible to concentrate pectin extracts to higher than a few per cent pectin which means they are expensive to transport. Processes were therefore developed to precipitate pectin in solid form. It then became possible to exploit the large quantities of citrus waste produced by the citrus juice industry, because the precipitation processes eliminated the bitter materials present in liquid citrus pectin extracts.

Up to the present time, these two sources, apple pomace and citrus peel, have provided the raw materials for the commercial pectin industry. The amount of suitable apple pomace is currently decreasing, as the juice industry turns to the use of pectolytic enzymes to increase yields. However, there are ample quantities of citrus peel available on a global basis, although the quality can vary considerably, and there is competition to obtain peel of the best pectin quality.

Other by-products from the food industry which contain significant quantities of pectins include sugar beet pulp and sunflower heads. Beet pulp is produced by the sugar industry in temperate climates on a vast scale, and most is used for cattle feed, although some added-value products which exploit the dietary fibre content of the pulp are being developed. Beet had been used in times of crisis to supplement apple as a source of pectin, but the acetylation of the pectin, and its relatively low molecular weight, make it a poor product for most applications. It is possible to devise chemical processes to reduce the acetylation while controlling the degree of esterification, but in common with all processing of pectin there is a loss of molecular weight. The final pectin produced has very low molecular weight and therefore very poor gelation properties. There is, however, a potential use for beet pectin in which the acetylation has been carefully retained. This arises from the long-known fact that beet pectin is more surface active, and produces troublesome foam during extraction. It has now been demonstrated that this pectin also has useful food emulsifier properties, and can be used to stabilise oil–water emulsions.

There are two major problems with sunflowers as a source of pectin. One is that at commercial maturity, from the point of oil production, the quality of the pectin has declined well below the optimum — this is in fact beneficial because the seeds are easier to separate from the heads. The other is that it is very difficult to avoid contamination of the pectin with traces of highly unsaturated oil. This oil, being finely dispersed, is very easily oxidised, leading to rancid off-flavours. There is again a problem with acetylation, but the potential molecular weight of the pectin is higher

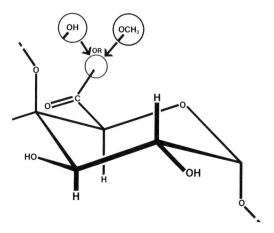

Figure 6.1 Galacturonic acid unit from a pectin molecule.

than with beet, so it is more feasible to deacetylate and retain adequate gelling performance.

6.2 Chemical nature of pectin

For a long time, the chemistry of pectins was poorly understood, and even today not all aspects are clear. Pectins are in fact the partial breakdown products of complex structures in the plant cell wall, and as such are invariably heterogeneous. Unlike a typical microbial polysaccharide, it is by no means clear whether there is a definite oligosaccharide repeat unit even within the pectin prior to extraction, and it appears that side chains and ester groups are lost to a greater or lesser degree not only during extraction of the pectin, but during the ripening process of the fruit. Ripening is clearly associated with a partial breakdown of the pectin within the cell wall, and this is largely responsible for the softening which occurs.

The characteristic chain-building unit of pectic substances is galacturonic acid, 1,4-linked into the polysaccharide chain (Figure 6.1). Many of these units are esterified with methanol, but the ester groups may be easily removed by the action either of enzymes in the tissue or of enzymes from yeasts and moulds. The main chain also contains occasional rhamnose units. Side chains containing mainly arabinose and galactose occur to a greater or lesser degree. The linkages of these side chains are more or less labile, and much neutral sugar material is lost in commercial pectin extraction. Commercial pectins show a distribution in both molecular

weight and degree of esterification, and these distributions will influence
the properties as well as making characterisation more difficult.

6.3 Pectin manufacture

Pectin production depends on a number of factors, including access to
raw material, water, energy and effluent disposal at reasonable prices. In
contrast to many other hydrocolloids, no crop is grown or harvested with
the primary aim of producing pectin. Like gelatin, pectin is produced
essentially from waste materials, and is dependent on the operations of
the primary industry of juice production. Hence the availability of raw
material, and its quality, are largely dependent on the market for various
types and qualities of juice and, to a lesser extent, essential oils. These
markets determine in the long term which fruits are grown, and in the
shorter term which fruits are purchased for processing, and at what
maturity. In some areas, the fresh fruit trade is the major market for the
fruit, and only surplus or low quality fruit reaches the juice factory. In
other areas, such as Brazil and to a degree Florida, large quantities of
fruit are grown specifically for juice. With apples, only the traditional
cider varieties are grown specifically for juice production, but these are
fortunately very suitable for pectin manufacture.

Apple pomace of good quality is obtained only from fresh rather than
cold-stored fruit, and is therefore very seasonal. Wet apple pomace is in
any case extremely difficult to process, so all pomace for pectin production is dried and stored until required. The citrus season can be much
longer, and in some areas, such as Brazil, citrus fruit of one sort or
another can be available for much of the year. Under these conditions the
peel can be used directly, but it can equally be dried and if necessary
transported long distances.

The present structure of the industry has been created by a combination of these factors. In the US, apple pectin producers had higher
production costs than the citrus pectin producers in California, and the
market became dominated by citrus pectin. More recently, environmental
restrictions in California have caused the closure of pectin production
because of the cost both of effluent treatment and of water supplies. In
contrast, European pectin plants, founded to produce apple pectin, have
been able to expand by importing dried citrus peel from a variety of
sources. This has given a degree of protection against poor harvests in
any one area.

The latest development has been the involvement of European com-
panies in production in major citrus-growing areas, such as Brazil and
Mexico. More investment in citrus-growing areas either inside or beyond
the European Community (EC) is likely in future.

The first essential of pectin manufacture is to ensure the best quality and treatment of the raw material. As soon as the fruit is processed, the pectin in the residue is under attack both by enzymes liberated within the fruit tissue and by those produced by yeasts and moulds. It is important to convert the highly perishable wet residue either into a pectin extract or into a more stable dry form as quickly as possible. With citrus peels, it is also vital to wash out the acid, which otherwise becomes concentrated during drying and causes further degradation of the pectin.

To extract the pectin, wet or dry peel or dry pomace is heated under carefully controlled conditions of pH and temperature in hot water containing mineral acid. It is necessary to use a large volume of liquor, as one of the difficult stages is the separation of the pectin extract from the remaining solids. If apple pectin is being produced, especially to be sold in liquid form, the extract is treated with carbon to remove some of the colour, and then with amylase to destroy the starch present. Further filtration removes small particles of cell wall and insoluble impurities, giving a reasonably bright extract.

Most pectin is now isolated by alcohol precipitation. For this to be efficient, the pectin extract must be concentrated. To avoid thermal degradation of the pectin, this is carried out in vacuum evaporators. The extract can then be either mixed directly with a suitable alcohol, which may be isopropanol, ethanol or methanol, depending on local costs and availability, or processed to modify the degree of esterification before precipitation. The fibrous or somewhat gelatinous precipitate is separated, and may then be subjected to further modification treatment before being dried, ground and blended into a uniform batch. This batch is then analysed for purity and gelling characteristics. Most pectin is sold diluted with sugar to a standard level of performance, in terms of either gel strength or some more relevant functional property.

6.4 Modification of pectin

Pectin as first extracted has a relatively high degree of esterification, around 70–75% of the acid groups in the molecule being naturally esterified with methanol. Such a pectin is ideal for use in a conventional jam, and will give a rapid set to prevent the floating of fruit. However, other uses require the preparation of pectins with different setting characteristics (Figure 6.2). Pectin is modified by reducing the degree of esterification (commonly referred to as DM or degree of methylation). This is most commonly carried out by acid hydrolysis.

This can be carried out at several stages of the process, in either the raw material, the extraction process, the concentrated extract or the wet precipitated pectin. Alternative processes could be to use alkali at low

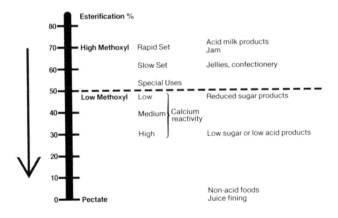

Figure 6.2 De-esterification of pectin to give a range of products.

temperatures, purified pectin esterase or ammonia. If ammonia is used, some of the ester groups are converted to acid amide groups ($-CONH_2$), producing an amidated pectin.

If the DM is reduced to around 60%, the pectin is a slow-set type, and is capable of gelation only in high-sugar systems. Under similar conditions, the gel will set more slowly or at a lower temperature than with the original rapid-set pectin. However, it also will tolerate either higher sugar concentrations or a lower pH. Once the DM is below 50%, the pectins are known as low methoxyl, and become steadily more reactive with calcium. They can be gelled under progressively lower soluble solids conditions, provided an appropriate amount of available calcium is present. Amidated pectins are mostly of the low methoxyl type, and have the advantage of tolerating more variation in calcium content.

6.5 Properties of pectins

6.5.1 Solubility and viscosity

Pectin is soluble in cold water and gives a viscous solution. Like other viscous gums, it needs care in dispersing the powder rapidly into the water; lumps of powder easily become coated with a gel layer, which makes further dissolving slow and difficult. It is important therefore to adopt a technique which facilitates dispersion to the individual particle level. This can be achieved either using a high-shear mixer, or by diluting the pectin with a soluble solid such as sugar. Because pectin is only very slowly soluble in concentrated sugar solutions, it is possible to disperse the powder easily into a syrup (high-fructose syrups are especially useful because of their low viscosity) and then to dilute this with water to below 20% total solids, when the pectin will dissolve on heating with only gentle stirring.

Although pectin is viscous in solution, it is not the most viscous of gums, and it would not usually be chosen where an increase in viscosity is the only desired effect. However, where only a small increase in viscosity is required, for example to replace the mouthfeel of sugar in a low-calorie drink, pectin can prove very effective.

6.5.2 Degradation

The pectin molecule is quite easily degraded. Although pectin, in contrast to some other gums, is fairly stable under acid conditions, its chemical structure makes it prone to breakdown under less acid conditions, at a pH of 5 or above, especially at higher temperatures. Because the reaction can occur wherever there is an esterified acid group in the molecular chain, a small amount of degradation can give a large loss in viscosity, gelling power and other functional properties. The rate of degradation is less with low methoxyl pectins, and virtually absent in pectic acid, the fully de-esterified material, and its salts. Hence, only pectic acid or pectins of very low degrees of esterification can be used in neutral products, especially those which are to be heat processed.

6.6 Gelation: high methoxyl pectins

High methoxyl pectin is best known as a gelling agent. However, it will gel only under rather specific conditions of sugar concentration and acidity, in which the pectin chains are partially dehydrated and the negative charge sufficiently reduced to permit chain–chain interactions. Typical relationships between gel strength and pH are shown in Figure 6.3. At a given sugar concentration, in this case 65% added as sucrose (of which a certain proportion will have been inverted to fructose and dextrose during the preparation of the gel by boiling), slow- and rapid-set pectins show a maximum pH of gelation of around 3.2 and 3.4 respectively. At these values are approached, the temperature at which a set occurs decreases towards ambient, and the final gel strength decreases progressively. Conversely, at lower pH values, the strength reaches a plateau and the setting temperature increases.

The optimum conditions for gelation are influenced by a number of factors. At higher sugar concentrations, the curves shown in Figure 6.3

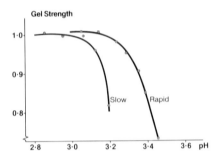

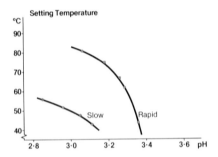

Figure 6.3 Effect of pH on high methoxyl pectin gelation.

move to a higher pH, and in sugar confectionery manufacture it is usual to use a slow-set pectin at a pH of between 3.4 and 3.8 depending on the sugar content. The minimum sugar concentration at which high methoxyl pectin will give a gel is between 50 and 55% with a very rapid-setting pectin at the lowest acceptable pH (around 2.8). Even at 60% solids, only a rapid-set pectin will gel at the normal jam pH of 3.0–3.2 and because of the effect on the setting–temperature curve, setting is relatively slow.

The nature of the sugar also has a marked effect on the setting rate. Replacement of some of the sucrose by glucose syrup leads to a decrease in gel strength, but an increase in setting temperature. This limits the degree of substitution of sucrose by the cheaper syrup. At high levels of glucose syrup there is also a tendency to obtain a pasty, starchy texture. The use of high-fructose corn syrup has less effect on gel strength, but causes a drop in setting temperature, especially with rapid-set pectin. It is therefore common in the US, where both types of syrup are readily available and cheaper than sucrose, to use a mixture of the two with perhaps a small amount of sucrose. The gelation properties in this system are much closer to those of sucrose, but should not be assumed to be identical. A curious fact is that maltose shows very similar effects to a 42

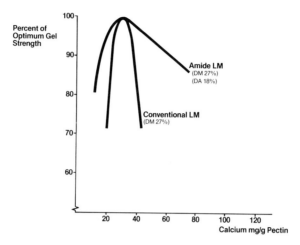

Figure 6.4 Effect of calcium level on low-methoxyl pectin gel strength.

DE glucose syrup, and similar effects should therefore be expected with high-maltose syrups.

6.7 Gelation: low methoxyl pectins

Low methoxyl pectins, whether or not they are amidated, form gels by reaction with calcium ions. They are believed to gel by the 'egg box mechanism' (Grant *et al.*, 1973) first suggested for alginates. A section of two pectin chains, which must be free from ester groups but may contain a limited number of amide groups, is held together by calcium ions. Although the amide groups can be incorporated into such a junction zone, they only contribute a little to the strength of the bonding between the chains. The major advantage with amidated pectins is their greater tolerance to changes in the calcium level (Figure 6.4). By controlling the proportion of acid and, if present, amide groups, a range of pectins with widely varying reactivites to calcium can be obtained. In the limit, pectic acid and its salts show a very high reactivity to calcium (Buhl, 1990), exceeding that of alginate, but in practice it has found as yet relatively few applications. In contrast, pectins of DM around 45% require 50 or 60% soluble solids in addition to calcium to form a gel.

In all low methoxyl pectin systems, gelation is governed by a number of parameters: calcium level, sequestrant level (which controls the availability of the calcium present), pH, soluble solids and the inherent reactivity of the pectin. These factors affect both the strength of gel produced by a standard quantity of pectin and its setting temperature.

With low methoxyl pectins, in contrast to high methoxyl gelation where time dependency is the rule, gelation is a function of temperature, and a unique setting temperature can be measured for any particular composition. The usual standard assessments of low methoxyl pectins are made at pH 3.0 and 31% soluble solids (30% added sugar in the form of sucrose), in line with the 'Exchange' method for grading (Food Chemicals Codex, 1972). It is preferable to modify the method so that the total citrate level is constant, and any required pH adjustment is made with a non-calcium-complexing mineral acid or alkali.

In a given system, containing, for example, 1% standardised commercial pectin (between 0.5 and 0.8% pure pectin, depending on type and quality), no gel will be formed below a certain minimum calcium content in the standard formulation. Above this level, the gel strength will increase rapidly with increasing calcium content, reach a maximum and then decline. Over this range, the setting temperature will increase from near ambient and approach boiling point as the maximum gel strength is reached. The reason for the decline in gel strength is that in a conventional hot-prepared jelly gelation actually starts to occur during the preparation of the gel mixture and this part of the strength is not recovered after pouring. Such gels are also very brittle, cloudy and prone to syneresis. To obtain non-brittle, elastic gels without syneresis it is essential to work below the maximum gel strength and setting temperature, in other words at lower than the apparent optimum calcium content.

The level of sequestrant is also crucial. As would be expected, if sequestration is increased, the system gels less readily and at a lower temperature. If sequestrant levels are reduced, gel strengths and setting temperatures are increased, and with no sequestrant it can become impossible to form a gel without a preset, broken texture.

The role of pH is more complex, but in general as pH is increased the tendency to form a gel is somewhat reduced. For example, to gel a sweetened but not acidifed milk system which has a high level of calcium, a high reactivity pectin would normally be used. At lower pH, the same calcium level would require a less reactive pectin.

As the water activity of the system is reduced, increasing the soluble solids of the system by including more sugar, it is easier to form a gel, and less calcium, or a less reactive pectin, is required. If low methoxyl pectins are to be used at soluble solids of above around 60%, it becomes essential to strongly sequester calcium in the system to balance this effect of increasing solids.

Although this may all sound rather daunting, in practice these effects make low methoxyl pectin systems very versatile and controllable in the hands of a technologist who has an understanding of the basic properties and experience of handling them and many attractive products can be produced.

6.8 Interaction with other polymers

Pectin is a negatively charged hydrocolloid and as such interacts with positively charged polymers in solution. The interaction with proteins below their isoelectric point has been studied (Glahn, 1982) and is the basis of the stabilisation of acid milk and soya milk systems by pectin. Pectin also interacts with gelatin at low pH, and it is important when using mixed pectin–gelatin gelling systems to control the pH and the ionic strength to prevent an unfavourable interaction.

A quite different type of interaction can be observed between high methoxyl pectin and high guluronic acid alginate. An approximately equal mixture of these will produce a gel in which neither will gel separately, at low levels of total gelling agent and sugar solids. However, this interaction is very sensitive to pH, and interrupted by quite low levels of calcium ions. Only a limited number of food products have used this interaction, perhaps because of the widespread occurrence of calcium in foods and the low melting point of the gels produced.

6.9 Physiological properties: dietary fibre

The major physiological properties of pectin, as far as ingestion is concerned, come into the realm of 'dietary fibre'. The whole subject area is complex and incompletely understood at present (British Nutrition Foundation, 1990; Southgate, 1990). Pectin is a soluble non-starch polysaccharide which is not digested by human gut enzymes, and thus it clearly falls within the definition of a soluble dietary fibre. It appears to have beneficial effects on cholesterol levels if taken in sufficient quantity, and it may have other advantages. However, because of its functional properties, it is a challenge to incorporate high levels of pectin into manufactured foods.

6.10 Legal status

Commercial pectin is regulated as a food additive in most countries, and is defined as such by the Codex Alimentarius Commission of the United Nations, by the EC and in the US. All these bodies issue criteria of purity from time to time, and their literature should be consulted for the latest versions. In general, pectin is well regarded from a toxicological point of view, and both pectin and amidated pectin (collectively 'pectins') have been granted an acceptable daily intake value (ADI) 'not specified' by the international expert committees JECFA (Joint Food and Agriculture

Organization World Health Organization Experts Committee on Food Additives) and the EC Scientific Committee for Food.

In terms of usage, pectin is widely permitted in a range of food types, although individual country regulations vary and should be consulted. Most quantitative restrictions arise from a desire to avoid any risk of consumer deception, for example the possibility of hiding a deficiency in the quantity or quality of fruit in a product, rather than from any toxicological concern. However, it should not be assumed without further enquiry that the use of a high level of pectin for dietary rather than functional purposes is covered by authorisation for use as a food additive.

6.11 Jams and jellies: traditional high-sugar products

Jam making is a complex operation involving a number of variables. An excellent overall review is given by Broomfield (1988). The description given here will concentrate on the role of pectin. Jams, jellies and marmalades are mixtures of sugars and fruit or fruit components, brought to a thickened or gelled consistency, usually by boiling. The boiling process ensures the destruction of fruit enzymes, extracts some of the pectin from the fruit and concentrates the product to a point where, as a result of its acidity and reduced water activity, it is self-preserving. This normally means a total solids content measured by refractometry on the sucrose scale of 63–70%. When this process is performed in the traditional way, without any further additions, the consistency that is obtained will vary considerably, depending on both the nature of the fruit and other factors such as variety and ripeness. At the high fruit levels of a traditional domestic jam, the consistency can vary from a very firm set to a sauce-like thickened product, or even thinner! It was against this background that the early use of pectin developed, and jam remains the largest single application for commercial pectin. In these traditional jams, there is almost always some contribution to the texture from the pectin extracted from the fruit. The added pectin level is therefore adjusted to allow for the fruit type and content, and Table 6.1, which covers a broader range than the legal fruit contents, gives typical pectin addition levels for a range of fruits. It is always wise, however, to carry out a small-scale trial on any new recipe, or fruit source, to adjust to the exact texture required.

The choice of pectin type depends on both the type of product and the process. In jams containing discrete fruit pieces, a major consideration is good distribution of fruit in the final pack. This is particularly critical in whole strawberry and even more in cherry jams. It is best in this case to use a fairly rapid-setting pectin and to control the filling temperature carefully so the fruit does not have time to float before some set has

Table 6.1 Pectin levels for traditional jams (in g/100 kg).

Jam type	Fruit content (%)			
	30	40	50	60
Cherry Peach Pear Pineapple Raspberry Strawberry	360–450	260–330	180–230	110–160
Apricot Blackberry Blackcurrant Marmalade	270–330	190–250	120–170	70–110
Apple Damson Gooseberry Greengage Guava Plum Quince Redcurrant Marmalade	180–240	110–170	60–110	30–70

developed. At the other extreme, a clear juice-based jelly must be free from retained bubbles, and a slow-setting pectin is essential. One of the more difficult preserves from this point of view is jelly marmalade, a clear citrus juice-based jelly with a few shreds of peel. The usual answer here is to use a medium rapid-set pectin, or a mixture of rapid and slow set, with careful process control, including perhaps treatment of the peel shreds before adding them to the product. The shreds in these products are usually very thin and so equilibrate with the sugar content and density of the bulk phase fairly quickly.

There are two major methods of jam manufacture: the traditional open pan system, usually but not always conducted in batch steam-heated pans, and the vacuum boiling process, which enables the product to be cooked at a lower temperature and to retain a fresher, less caramelised, flavour and colour. This can be carried out in a range of types of batch cooker, or in continuous evaporators suitable for handling the particle size of the fruit involved. In either case pectin must always be added as a solution free from any undissolved material. This is usually added quite late in the process to reduce the degradation of the pectin during the heat process. With vacuum boiling, the usual technique is to boil under vacuum to a higher solids value than the final product, to break the vacuum and raise the temperature to at least 85–90°C, and to add the pectin at this temperature. The product is then completed by adding the appropriate quantity of acid solution to give the required pH and water to adjust the solids. As discussed in the section on the gelation of high methoxyl pectin, the control of these two parameters is critical to control the texture of the jam.

Once the acid has been added, the jam should be filled into its final containers without delay, as the setting process is dependent on both time and temperature. If the jam is not filled before setting starts, a 'broken' texture which is much more prone to eventual syneresis will result, although this can be hidden by the texture of the fruit pulp to some extent. In extreme cases, the gel may be so sheared during transfer and filling operations that a slightly thick but apparently unset product is obtained. There is then a temptation to make the next batch with more pectin. Increasing the pectin level will increase the rate of set, making the problem worse rather than better.

The main causes of problems in making preserves are:

1 Using pectin which is not completely dissolved, or stale, degraded pectin solution.
2 Adding the pectin solution under conditions where gelation would be expected, e.g. at too low a temperature or pH.
3 Poor control of the final sugar solids or pH.
4 Filling at too low a temperature for the formulation, or after an excessive delay.

To these should be added the failure to allow for the changes in setting behaviour if the nature of the sweetening sugars or syrups is changed. If all these factors are observed, and found to be correct, then it is possible that there has been some change in the contribution from the fruit pectin, owing to the age, processing or variety of the fruit pulp.

A typical recipe for a strawberry jam, which can easily be adapted to higher or lower fruit content, is as follows:

Ingredients
(A) Sucrose	610 g
Frozen strawberries	400 g
Rapid-set pectin 4% (w/v) solution	75 g
(B) Citric acid monohydrate 50% (w/v)	6 ml
Final batch weight	1 kg
pH (as is)	3.0–3.1
Soluble solids	65–66%

Method
1 Prepare the pectin solution in (A).
2 Add ingredients (A), i.e. sucrose, fruit and pectin solution to a saucepan.
3 Heat the contents to boiling.

4 Boil the weight down to 1015 g.
5 Add the acid in (B), cool with stirring to 85°C.
6 Deposit into jars.

Notes

1 Fruit contains approximately 10% soluble solids. Therefore, if altering the quantity of fruit added, the amount of sugar must also be altered to take account of this.
2 The acid must be added at the end of the boil to prevent the mix being too viscous during the boil and to avoid pregelation.
3 The level of acid must be altered depending on the type of fruit used to maintain the pH at 3.0–3.1.
4 The pectin level may be adjusted to give a firmer or weaker set. Also, if adjusting the fruit level, the pectin extracted from the fruit must be taken into account.

6.12 Jams and jellies: reduced sugar

With increasing concern about total energy intake from food, and especially from sugars, the idea of reduced-sugar products has become more appealing. Unfortunately, the conventional high methoxyl pectin gelling system, making use of the pectin from the fruit in addition to added pectin, fails entirely below about 55% sugar solids on the refractometer scale. It is therefore necessary to find another gelling system, and to use this to provide virtually all the texture of the finished preserve. Fortunately, low methoxyl pectins of both conventional and amidated types have been developed for exactly this situation. Because they gel by reaction with calcium either naturally present or added to the system, they can be used down to around 10% total solids, although increasing care is required as the sugar content is reduced.

The general indicators for the choice of low methoxyl pectin type have been discussed in the section on gelation. It is again most important to ensure that the pectin is correctly dissolved, although amidated low methoxyl pectins, perhaps because of their reversible gelation, can be more forgiving in this respect. Especially at the lower sugar contents, recipes can be formulated which require little or no water removal by evaporation. However, it is still necessary to ensure sufficient heat treatment to eliminate yeasts and moulds and to deactivate fruit enzymes. Even traces of pectic enzymes can result in products which will mysteriously liquefy during storage.

With low methoxyl pectins, preserves can be made over a wider range of pH values, certainly between 3 and 4. The selected pH value and the addition of fruit acids or their salts to control it will have an impact on the reaction between the pectin and the calcium required to gel it.

The calcium content of the system has a crucial role in determining the strength and rate of set of the gel. However, what is important is not the total amount of calcium but the amount of calcium available to the pectin. As this is determined by the competition for calcium between the pectin and the various other sequestrants present, including fruit acids, sugars, sugar alcohols, etc., the available calcium cannot be determined analytically, and it is essential to perform a series of trials. It is worth noting that at moderate sugar solids and a pH around 3 the presence of some sequestrant is usually required to produce a gel without pregelation. This is not normally a problem in jams, as enough is provided by the fruit.

The setting temperature of low methoxyl pectin gels is determined entirely by their composition, and is not a function of time, and it is essential to fill low-sugar jams only just above the setting temperature, as the low viscosity makes fruit separation much more probable.

In products where sugar is entirely replaced by concentrated fruit juices, there are special problems of calcium availability because of the high concentrations of fruit acids and their salts. It is essential to use a more calcium-reactive pectin, and it is often desirable to add further calcium as a weak solution of a soluble salt such as calcium lactate or chloride.

6.13 Industrial fruit products

6.13.1 Fillings

It is difficult to give precise recommendations for the preparation of bakery jams because formulation and procedures depend very much both on the manufacturing plant available and on the exact conditions under which the jam or filling is to be used. Fruit content and state (pulp, purée, etc.) can vary, and some products will use an inexpensive fruit such as apple or plum pulp as a basis, with added flavouring and colour. It is common to use higher levels of glucose syrups than in retail jams, both for cost reasons and to give the heavy thickened texture which is often required. In some applications, texture and freedom from syneresis, and hence a reduced tendency to soak into the baked product, is the main requirement. In other cases the jam must resist baking. In this case a medium rapid-set pectin which will give the best balance between slow setting of the bulk jam in manufacture and the tendency to boil out is normally recommended. A recent requirement is to produce a bakefast jam at reduced sugar content, less sweet and lower in calories, as in the following recipe:

Heat-stable jam at 50% soluble solids

A jam can be produced that will withstand baking temperatures of 240°C for up to 12 minutes.

Ingredients

(A) Fruit		450 g
Water		50 g
Citric acid monohydrate 50% (w/v)		4.0 ml
(B) Low methoxyl pectin		8 g
Caster sugar		60 g
(C) Granulated sugar		300 g
42 DE glucose syrup (82% soluble solids, SS)		100 g
Final batch weight		1 kg
pH (as is)		3.1–3.3
Soluble solids		50%

Method

1 Warm ingredients (A), add ingredients (B) (premixed) and heat to boil.
2 Boil for 2–3 min.
3 Add ingredients (C).
4 Boil to 1 kg.
5 Cool with stirring and deposit at 50°C.

Notes

1 Adjust the pectin level to obtain the consistency required.
2 Depending on the type of fruit used it may be necessary to alter the pectin type or to add calcium to obtain bakefastness.

6.13.2 Fruit bases

As the market for fruit yogurts and other dairy desserts containing fruit has grown, a wider range and higher quality of fruit bases have been demanded. There has been a tendency to move away from dependence on modified starches as thickeners, and towards pectin and other materials which can give better texture and taste in the final product, and have a better marketing image. Similar considerations apply to fruit toppings for cheesecakes and similar desserts. The variety which can be produced makes it difficult to quote a truly typical recipe, as fruit and sugar contents can very widely, and the product must retain fruit suspension while being both pumpable and easily mixable where bases for incorporation into 'Swiss-style' yogurt are concerned. Most use either a conventional or an amidated low methoxyl pectin, as in the following example:

Fruit base

This is a fruit preparation which is thickened with a conventional low methoxyl pectin or an amidated low methoxyl pectin, giving sufficient structure to prevent fruit separation during storage, transport or use.

Ingredients

(A) Low methoxyl pectin   9 g
    or amidated low methoxyl pectin   8 g
    Sugar   40 g
    Water   200 g

(B) Strawberries   500 g
    Sugar   210 g
    Water   70 g
    Citric acid monohydrate 50% (w/v)   As required
    Sodium citrate dihydrate 20% (w/v)   As required

(C) Calcium lactate pentahydrate 3% (w/v)   As required

    Final batch weight   1 kg
    pH   3.5
    Soluble solids   30%
    Filling temperature   30°C

Method

1 Dry mix pectin and sugar. Dissolve in water using a high-shear mixer. To ease dissolving the water may be heated to 60°C.
2 Weigh ingredients (B) and the pectin solution into a pan and heat to 90°C.
3 Add part (C) if required and maintain temperature at 90°C for 10 min to pasteurise, or heat until the batch weight has been reduced to 1015 g.
4 Cool the fruit preparation while stirring to 30°C to obtain a thickened rather than a gelled texture.

Notes

1 If the viscosity of the base is not sufficient, calcium may be added in the form of calcium lactate or chloride. This should be added at the end of the boil as a dilute solution.
2 The pectin level may be altered to obtain the degree of thickening desired.
3 Citric acid or sodium citrate may be added to alter the pH of the product. In doing so it may also be necessary to alter the pectin level or type.
4 The pectin level and type may also be altered to obtain the setting or filling temperature suitable for the processing equipment available.

6.13.3 Glazes

An interesting application of the reversible gelation properties of amidated low methoxyl pectins is in the production of glazes for bakery products, both fancy pastries and fruit flans and the like. There is a market for a shelf-stable glaze base which can be used by the craft baker as required to impart a glossy glazed finish which is not too heavy in texture or too sweet. This can be accomplished by making use of the strong calcium sequestering power of food-grade diphosphates. Pectin manufacturers can supply blends of the required ingredients to produce glazes either with or without the inclusion of fruit pulp, as in the following recipe:

Flan-glazing jelly

This is a flan-glazing jelly containing fruit pulp which is produced using a buffered amidated low methoxyl pectin and which is fully heat reversible and may be further diluted with water.

Ingredients

(A)	Water	190 g
	Buffered amidated low methoxyl pectin	10 g
(B)	Sieved apricot pulp	100 g
	Water	100 g
	Citric acid monohydrate 50% (w/v)	2 ml
	Sugar	500 g
	42 DE glucose syrup (82% SS)	200 g
(C)	Colour and flavour	As required
	Final batch weight	1 kg
	pH	3.5–3.7
	Soluble solids	69%

Method

1 Using a suitable high-speed mixer, dissolve pectin in water (A) at 80–85°C. This hot solution should be used within 15–20 min of preparation.
2 Heat together ingredients (B) and boil down to 75–76% soluble solids.
3 Stirring thoroughly, pour the hot pectin solution into the saucepan.
4 Boil to 1010 g and add colour and flavour.
5 Stirring slowly, cool to at least 40–45°C.

Application

Liquefy the glazing product either by itself or diluted with up to one-half of its weight of water by heating to about 85°C. This may then be spread on the fruit or flan at any temperature down to about 60°C. Any unused glazing product may be liquefied and set back several times without loss of quality.

Notes

1 This pectin blend has been specially developed for use in glazes containing fruit pulp. Buffered amidated low methoxyl pectins may be used in products containing no fruit.
2 To increase the strength or dilutability of the glaze, the pectin level may be increased.

For industrial bakery use, it may be more convenient to make up the diluted form of the glaze directly. If necessary, this can still be remelted for use on a later occasion, although at the lower soluble solids it does not have the same microbiological stability as the concentrate.

An alternative glazing system uses the cold-setting properties of high methoxyl pectins. High methoxyl pectin can be formulated into a syrup at a pH sufficiently high to prevent gelation, but which is still acidic enough to be stable at the high sugar concentration used. Addition of a small amount of citric or other fruit acid will cause gelation to occur after a few minutes, sufficient time to apply the syrup to fruit flans or other bakers' wares.

Cold-setting flan jelly

This is a syrup produced using a buffered rapid-set pectin which, upon addition at room temperature of a controlled amount of acid, will set within a few minutes and provide a glazing for fruit tarts or flans.

Ingredients

(A)	Water	400 g
	Buffered rapid-set pectin	7 g
	Sugar	30 g
(B)	Sugar	470 g
	42 DE glucose syrup (82% SS)	210 g
(C)	Colour and flavour	As required
	Final batch weight	1 kg
	pH	4.0
	Soluble solids	68%

Method

1 Mix pectin and sugar (A) thoroughly in the dry state and gradually add to the water which has previously been heated to about 70°C.
2 Stir continuously and bring mixture to the boil. Boil for 1 min to ensure that all the pectin is dissolved.
3 Gradually add sugar (B), and when fully dissolved add the glucose syrup.
4 Boil rapidly to 1000 g.
5 Mix in colour and flavour. Fill into containers and cool as soon as possible.

Application

One kilogram of this syrup can be made to set at room temperature in approximately 3 min by the addition, with rapid stirring, of 4.4 ml of citric acid monohydrate 50% (w/v). A longer setting time, approximately 10 min, can be obtained by halving the quantity of citric acid.

6.14 Confectionery

The major use of pectin in the confectionery industry is to prepare light-textured jellies with either acid fruit or more neutral flavours. These products also include chocolate-enrobed jellies, fruit slices and cake decorations. Confectionery jellies are almost always produced at 75–85% soluble solids to ensure microbiological stability, and this requires the use of a suitable sugar mixture, normally including some type of glucose syrup in the recipe. Standard recipes incorporate the long-established 42 DE grade of syrup, but other types are now available. They may have slightly different effects on the setting behaviour of pectin, and if any difficulties are experienced the pectin supplier should be asked for advice.

Acidic or fruit-flavoured jellies are produced to a pH of 3.3–3.8, depending on the soluble solids required. A slow-set pectin is normally necessary to prevent presetting. It is essential to control the pH of the mixture during cooking, especially in traditional batch open pan equipment, and suitable buffer salts are an essential ingredient. The type and quantity of buffer salt can be varied to give different perceived acidity in the final jelly. If fruit pulp is used, this will itself contribute some buffering, but it will still be necessary to control the pH to a sufficiently high, but not too high, pH of ideally 4.3–4.8. In this range, pectin is reasonably stable to degradation, but will not pregel. Just before depositing, the pH is reduced by adding a strong solution of citric or other fruit acid, and the mix must then be deposited within 20 min.

Confectionery jelly

This is a clear, clean-textured confectionery jelly produced using a slow-set, high methoxyl pectin and which requires no subsequent stoving.

Ingredients

(A)	Water	325 g
	Citric acid monohydrate 50% (w/v)	2.8 ml
	Potassium citrate monohydrate 40% (w/v)	6.25 ml
(B)	Slow-set pectin 150X SAG	8.8 g
	Sugar	40 g
(C)	Sugar	470 g
	42 DE glucose syrup (82% SS)	300 g
(D)	Citric acid monohydrate 50% (w/v)	6 ml
	Colour and flavour	As required
	Final batch weight	1 kg
	pH	3.3–3.6
	Soluble solids	76–78%

Method

1 Dissolve citric acid and potassium citrate in water (A), and heat to not more than 70°C.
2 Mix ingredients (B) in the dry state and gradually add to the water, stirring continuously.
3 Bring slowly to the boil while stirring. Boil for 1–2 min to ensure that all the pectin has dissolved.
4 Gradually add sugar (C) and, when dissolved, add glucose syrup.
5 Boil rapidly to 1000 g.
6 Immediately before depositing, mix in the citric acid thoroughly.
7 Mix in colour and flavour, and deposit into moulds. Provided that the temperature of the batch is kept above 90°C, depositing should be completed within 20 min of the final acid addition if premature gelation is to be avoided. (The depositing time should be slightly less if the water is hard.)

8 The final product should have a pH (50% solution by weight) in the range 3.3–3.6 and a soluble solids content of 76–78%.

It is also possible to modify the texture by adding gelatin, or a foaming protein as in the following recipe:

Aerated pectin confectionery jelly
This is an aerated pectin confectionery jelly produced using a slow-set, high methoxyl pectin and whipped using a dairy-based whipping agent.

Ingredients

(A)	Water	300 g
	Trisodium citrate dihydrate 20% (w/v)	5 ml
(B)	Slow-set pectin 150X SAG	8 g
	Sugar	80 g
(C)	Sugar	480 g
	42 DE glucose syrup (82% SS)	265 g
(D)	Dairy-based whipping agent	6 g
	Water	12 g
(E)	Citric acid monohydrate 60% (w/v)	3.6 ml
	Final batch weight	1 kg
	pH	3.6
	Soluble solids	78–79%

Method
1 Dry mix pectin and sugar in (B). Add (A) to pan, and while heating add dry mix with continual stirring.
2 Make sure all of the pectin has fully dissolved and then add (C) while maintaining the boil.
3 Boil contents down to 1 kg, take off the heat. Mix in (D) (which has previously been dissolved).
4 Beat the mixture for 2 min using a food processor, add the acid, then deposit.

Notes
1 The pH should be in the range 3.5–3.6; if not, the level of citric acid should be adjusted to achieve this.
2 Various textures can be achieved by altering the pH of the product. If a softer texture is required the pH can be increased to achieve this.
3 The acid is added during mixing to avoid the product setting before it has been deposited; to avoid this the pH can be slightly increased to achieve a lower setting temperature.

Jellies with neutral flavours may be produced using a low methoxyl pectin in conjunction with diphosphate buffer salts, which are neutral in taste and also control the availability of calcium to the pectin. Although it is possible to create a recipe from the basic ingredients, the procedure is complex, and it is better to use a premix of an appropriate pectin with the necessary buffer and calcium salts. A number of standard blends are available, but if process requirements are very specific a custom blend may be necessary. Either conventional or amidated low methoxyl pectins may be used, but amidated types are sometimes more successful in sophisticated large-scale plants. The following recipe provides a guide, but many variations are possible. Although the traditional Turkish Delight, with a rose flavour, is the most common product, other flavours such as peppermint, coffee, etc. are equally attractive.

Turkish delight jelly
This is produced using a buffered amidated low methoxyl pectin.

Ingredients

(A)	Buffered amidated low methoxyl pectin	15 g
	Sugar	60 g
	Water	300 ml
	Citric acid solution 50%	4 ml
(B)	Sugar	350 g
	42 DE glucose syrup (82% solids)	400 g
(C)	Colour and flavouring	As required
	pH (on a 10% solution)	4.6–4.8
	Soluble solids	75–77%
	Batch weight	1 kg

Method
Dry mix the pectin and sugar (A) and add to the water at 70°C (containing the acid if required). Stir vigorously to avoid lumping, and bring to the boil. Add the sugar and glucose (B) and return to the boil. Boil to the required solids by refractometer. Add colour and flavouring and deposit without delay. Either starch or continuous moulding techniques can be used. If an opaque appearance is required, up to 3% of a thin boiling starch may be added.

6.15 Desserts

With an increasing demand for variety in both convenience and luxury food items, the ready-to-eat dessert market has shown rapid growth in volume and variety in recent years. Pectin can be used in a variety of products: in glazings as described above, and in both water and milk gels of various textures. It can also be used for ice crystal control in sorbets and water ices. Use in milk systems will be discussed later.

Water gels made with pectin have several advantages over other gelling systems. The gels show good clarity and flavour release, and there is less restriction on the pH required to give good acidic flavours. The setting and melting temperatures can be controlled to facilitate processing and to avoid the melt-down that can occur with gelatin gels between the shop and the domestic refrigerator. However, if required, the gel can be melted and reset with minimal loss of strength. These gels can be produced over a range of soluble solids contents, and above 35% solids they are freeze–thaw stable and can be used in frozen desserts. They are very useful in multilayer situations, because the gelling behaviour can be precisely controlled. If a gel is deposited at its setting temperature, a further layer can be added to the dessert after a short interval, without elaborate cooling equipment. It is even possible to formulate a gel which can be pumped and will reset on depositing, as soon as the imposed shear has been removed.

A typical recipe is as follows:

Water jelly
This uses low methoxyl pectin.

Ingredients
(A)	Water	700 g
	Citric acid monohydrate	3.5 g
(B)	Amidated low methoxyl pectin	13 g
	or low methoxyl pectin	
	(conventional)	
	Sugar	250 g
(C)	Hot water	50 g
	Calcium lactate pentahydrate	2 g
	Final batch weight	1 kg
	pH	3.4
	Soluble solids	26–28%

Method
1 Dry mix the pectin and sugar (B); add this to the acid and water (A).
2 Heat to boiling, and add the calcium lactate solution (C) with stirring.
3 Reduce to 1 kg and deposit immediately, or cool to just above the setting temperature before depositing.

6.16 Cold-set desserts

It is quite possible to formulate dessert formulations which can use the cold-setting properties of low methoxyl pectins with calcium. A syrup prepared to the following recipe reacts rapidly with milk to produce an attractive lightly gelled dessert.

Milk fruit dessert
This is a fruit syrup which uses a low methoxyl pectin and which when mixed into milk produces a weak-gelled dessert within 5 minutes of mixing.

Ingredients

(A)	Low methoxyl pectin	20 g
	Sugar	80 g
(B)	Water	300 g
	Sodium citrate dihydrate 20% (w/v)	44.0 ml
(C)	Sugar	120 g
	Strawberries	600 g
(D)	Flavour	As required
	Final batch weight	1 kg
	pH	4.0–4.2
	Soluble solids	28–30%

Method

1 Dry mix the pectin and sugar in (A).
2 Weigh ingredients (B) into a pan, begin to heat and add dry mix. Heat to boiling while stirring.
3 Add sugar and fruit. Reduce to 1020 g. Remove from heat and add flavour.
4 Cool to 20°C.
5 To prepare dessert, take equal parts of fruit syrup and milk. Mix the milk into the fruit syrup for 25–30 s. The dessert may be eaten within 5 min of preparation.

Notes

1 When the milk is mixed into the fruit syrup the dessert will start to thicken up within 5–10 s. Continue mixing for 15–30 s until the mix is homogeneous.
2 The pH must be around 4.2. Adjust citrate levels as necessary.
3 A pectin solution may also be made using a high-shear mixer.
4 When using a different fruit variety the citrate level may need to be altered to achieve the desired pH.

6.17 Restructured fruit and vegetables

It is possible to use calcium-reactive low methoxyl pectin in much the same way as alginates for the production of reformed or restructured fruit and vegetable pieces. This approach depends on making up a mix either in the absence of calcium or under conditions under which the calcium is not free to react, and then either activating the calcium in the system by reducing the pH or diffusing calcium into the product from a high-calcium setting bath. Using these techniques, products ranging from imitation glacé cherries for bakery products, through fruit pieces, to restructured onion rings, can be produced. Many processes in this area have been patented, and it is wise to confirm the legal situation before producing a product of this type. A typical recipe is as follows:

Restructured onion

This is produced using an amidated low methoxyl pectin.

Ingredients

(A)	Amidated low methoxyl pectin	8 g
	Sodium polyphosphate (Calgon)	4 g
	Water	250 g
(B)	Dried minced onion	116.5 g
	Flour	100 g
	Salt	1 g
	Water	250 g
(C)	Calcium chloride hexahydrate	100 g
	Water	900 g
	Final batch weight	1 kg

Method

1 Dry mix ingredients (A) and dissolve in the water using a high-shear mixer.
2 Mix ingredients (B) together and leave for 30 min to allow the onion to rehydrate.
3 Mix (A) and (B) and extrude without delay into the calcium bath (C). This should be completed within 30 min of mixing to minimise the reaction between the pectin and the calcium in the onion.
4 Leave the extruded pieces in the setting bath for at least 10 min, then rinse to remove excess calcium chloride. The pieces can be fried, baked or frozen.

Restructured peach

This is a cold-set restructured fruit which uses an amidated low methoxyl pectin and can be used as a filling for bakery products, being bakefast and freeze–thaw stable.

Ingredients

(A)	Water (distilled)	380.5 g
	Sugar	100.0 g
	Amidated very low methoxyl pectin	15.0 g
	Dicalcium phosphate anhydrous	4.6 g
	Sodium citrate dihydrate	1.4 g
(B)	Peach purée	335.0 g
	Sugar	148.0 g
	Citric acid monohydrate	9.7 g
	Sodium citrate dihydrate	5.8 g
	Flavour	As required
	Final batch weight	1 kg
	pH	3.8–3.9
	Soluble solids	33–35%

Method

1 Dry blend ingredients (A). Dissolve in water using a high-shear mixer.
2 Mix ingredients (B) until sugar has dissolved.
3 Pour ingredients (A) into ingredients (B). Mix by hand until homogeneous (approximately 10 s) and pour into mould immediately.
4 Gel can be demoulded after 5 min.

Notes
1 If the product is setting too quickly, reduce the quantity of citric acid in (B).
2 This product is suitable for baking and will retain its identity when defrosted, unlike whole fruit.

6.18 Milk systems

6.18.1 Milk desserts

Because milk and milk products contain calcium, they can easily be gelled using low methoxyl pectins. It is possible to make both neutral milk desserts and gelled or thickened acid milk products such as yogurts.

Where permitted, yogurts may be thickened by adding between 0.05 and 0.1% low methoxyl pectin to the milk before culturing. It is necessary to disperse the pectin in the cold milk then heat to dissolve the pectin before starting the culturing process.

Fresh milk gels can be prepared by adding an amidated pectin. Depending on the reactivity of the pectin, textures from a firm gel to a creamy texture can be prepared. A set product can be prepared as follows:

Milk jelly
This uses an amidated low methoxyl pectin.

Ingredients
(A) Amidated very low methoxyl pectin	3.0 g
Sugar	4.5 g
(B) Fresh milk	500 ml
(C) Colour and flavour	As required

Method
1 Dry mix the pectin with sugar.
2 Warm the milk, and add the pectin–sugar mix (A) with stirring to ensure complete dispersion.
3 Continue to heat with stirring to the boil, and boil for 30 s.
4 Cool with stirring to 50°C, add colour and flavour, and pour into suitable containers.
5 Cool to 5°C.

Other softer textures can be obtained by using less reactive types of amidated pectin.

6.18.2 Cultured drinks

The market for acid milk drinks has increased markedly in a number of countries, partly because of the development of aseptic packaging techniques and the ability to use suitable stabilisers to enable heat-treated long-life products to be produced. Drinks based on yogurt are not new, but the ability to produce fresh-tasting sweetened drinks flavoured with fruit juices, etc. has broadened their appeal, especially when they can be made available in a shelf-stable form. Pectin has shown itself to be an ideal stabiliser for these products.

Typically, a level of 0.5–0.6% of a special pectin (for example, high methoxyl pectin) is added to the yogurt at 15–20°C. Fruit juice (or concentrate, flavouring, colour, etc.) is added, and the mixture homogenised at about 15 MPa after heating to around 40°C. The product can then be pasteurised at around 80°C and filled aseptically. It is advisable to carry out the heating gently, using minimum surface temperatures, such as a heat exchanger heated with hot water rather than direct steam to minimise the thermal shock to the product.

6.18.3 Whey drinks

Cultured or acidifed whey drinks suffer from protein precipitation if heat treated. Addition of pectin enables the production of shelf-stable aseptically filled products which can be very refreshing.

6.18.4 Milk/juice blends

Another variation is a direct blend of milk and fruit juice. Such products have poor stability even in a short shelf-life form without heat treatment. The addition of pectin to the milk before the fruit juice or other acid ingredient will result in a smoother texture with less tendency to give a sediment on standing. If a longer life product is required, it is essential to add pectin to avoid complete precipitation of the protein.

In all these products, careful control of the product composition, especially the pH, but also the salt content, is necessary. Typically, the best stabilising effect is obtained between pH 3.8 and 4.2.

6.19 Other food uses and potential uses

The applications of pectin have diversified considerably over the last decade. Although jams and jellies are still the major use, a wider range of desserts, stabilised acid milk products, and other products has been developed. Pectin has been used to improve the mouthfeel of low-calorie drinks, and as a thickener in savoury products such as salad dressings, sauces and spreads. There is also considerable interest in pectin as added soluble dietary fibre in foods and drinks. Some of these applications demand innovation from both the pectin supplier and the user. The unique properties of very low ester pectin (Buhl, 1990) have yet to be fully exploited, especially its stability to heat at neutral pH and its exceptional calcium reactivity. Further types of amidated and conventional low methoxyl pectins may well be developed and new techniques of carbohydrate chemistry may lead to a better understanding of pectin processing and applications. There is every reason to anticipate as much or more change in the next decade.

References

Braconnot, H. (1825) *Ann. Chim. Phys., Sér. 2*, **28**, 173.
British Nutrition Foundation (1990) *Complex Carbohydrates in Foods*. The Report of the British Nutrition Foundation's Task Force, BNF, London.
Broomfield, R.W. (1988) Preserves. In: *Food Industries Manual*, M.D. Ranken, ed., Blackie, Glasgow, pp. 335–355.
Buhl, S. (1990) Gelation of very low DE pectin. In: *Gums and Stabilisers for the Food Industry*, Vol. 5, G.O. Phillips, D.J. Wedlock and P.A. Williams, eds, IRL Press (at Oxford University Press), Oxford, pp. 233–241.
Food Chemicals Codex, 2nd edn (1972). National Academy of Sciences, Washington, pp. 580–581.
Glahn, P.E. (1982) *Progress in Food and Nutrition Science*, Vol. 6, G.O. Phillips, D.J. Wedlock and P.A. Williams, eds, Pergamon Press, Oxford, pp. 171–177.
Grant, G.T., Morris, E.R., Rees, D.A., Smith, P.J.C. and Thom, D. (1973) *FEBS Lett.*, **32**, 195.
Kertesz, Z.I. (1951) *The Pectic Substances*, Interscience Publishers, New York.
Southgate, D.A.T. (1990) Dietary fibre and health. In: *Dietary Fibre: Chemical and Biological Aspects*, The Royal Society of Chemistry, Cambridge.

7 Seed gums

J.E. FOX

7.1 Introduction

With a few exotic exceptions, life on earth depends on the energy
captured by photosynthesis in green plants and algae. Now, as every
holidaymaker will be aware, the sun does not shine on demand, particularly in the British Isles, and this applies not only to melanin-starved
holidaymakers but also to plants. Plants therefore need some mechanism
by which energy can be stored in chemical form. This is particularly
important where the plant is exposed to an inhospitable climate, such
as winter or drought, for prolonged periods and where relatively large
amounts of energy need to be stored to guarantee survival of the species.
In these conditions energy storage in the form of simple mono- or dis-
accharides has certain limitations. High concentrations of sugar can lead
to high osmotic pressures within the cell and high viscosities, both of
which the cell may be unable to tolerate.

These problems can be avoided if the energy carrier can be transferred
to a separate phase outside the aqueous cytoplasm. For this purpose
nature has availed itself of two pathways. Many plants and almost all
animals, including man, are able to convert excess sugars to water-
immiscible fats. The alternative is to polymerise the sugar in such a way
as to make the crystalline or solid phase thermodynamically more stable
than the solution at ambient temperatures. When energy demand is high,
these insoluble reserve polysaccharides can be enzymatically degraded to
their component soluble sugars and these metabolised as before. By far
the most common of such reserve polysaccharides is starch, a polymer of
D-glucose. However, many plants can polymerise other sugars to reserve
polysaccharides, and several of these have properties of interest to the
food technologist. To date, only one group of such polymers has been
commercially exploited, the galactomannans, and they are the subject of
this chapter.

7.2 Galactomannans

The seeds of many Leguminosae contain galactomannans in the cells of
the endosperm (Dea and Morrison, 1975). The polysaccharide is laid

Figure 7.1 Locust bean pods and seeds. The seeds are about 10 mm long and weigh approximately 0.25 g each, of which approximately 38% is galactomannan.

down not in discrete bodies within the cell, as is starch, but as thickening of the cell wall. The thickening progresses from the periphery to the cell centre, and in some species completely excludes the cytoplasm. The galactomannans from three species have been commercially exploited. They are those from locust bean, guar and tara. Those which are of interest but are not yet commercially available include *Cassia* species, mesquite (*Prosopis* species) and fenugreek (*Trigonella foenum-graecum*).

7.2.1 Locust bean gum

The locust bean or *Ceratonia siliqua* is indigenous to the Middle East and was cultivated by the early civilisations that flourished around the Mediterranean. The plant itself is a tree which grows to some 10 m in height and first bears fruit when some 10–15 years old. These fruits, which are about 10–20 cm long, make excellent animal feed, and this, together with the tree's ability to thrive on poor soil, accounted for its early popularity. Each fruit contains *c.* 10–15 seeds or carob beans, which are the source of the polysaccharide (Figure 7.1). These seeds are separated, the dark seed coat or testa roasted from the endosperm and, using techniques akin to those used in the flour industry, the endosperm is ground to the commercial locust bean gum. The standard gum is sold in a range of particle sizes of which the coarser gum powder has a slightly higher viscosity. The standard gum also contains finely ground pieces of the dark-brown testa, which appear as dark specks in the powder. Attempts to increase yields during grinding usually result in more specks, and this will detract from the quality of the product. A solution of locust bean gum is also somewhat milky owing to the presence of some un-dissolved fat and proteins, and improvement can be obtained by wash-

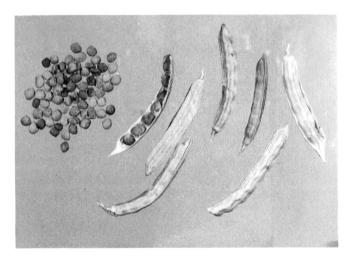

Figure 7.2 Guar pods and seeds. The seeds are about 4 mm in diameter and weigh
approximately 35 mg each, of which approximately 36% is galactomannan.

ing with alcohol. Where a completely clear product is required, as for
example in fruit jellies, alcohol-precipitated polysaccharide can be used,
but this is appreciably more expensive than the unrefined gum. As will be
discussed later, locust bean gum is only partially soluble in cold water and
to develop full functionality must be heated. So-called cold water-soluble
gums are available; these are either hot solutions dried in the presence of
materials, such as sugars, to prevent recrystallisation of the polysaccharide
or high-galactose fractions which have been selectively separated from the
crude gum.

7.2.2 Guar gum

Guar gum is obtained from the seeds of *Cyamopsis tetragonolobus*, which
is indigenous to NW India and Pakistan. In these countries it has been
grown for centuries both as an animal feed and for human consumption.
Its importance as a food additive grew during and after the Second World
War when supplies of carobs were restricted and cultivation of guar was
begun in Texas. Unlike carobs or tara, guar is an annual, being planted
after the monsoons in June/July and harvested in December. The plant is
c. 1.0 m high and carries green pods, somewhat smaller than those of the
domestic European pea, which contain up to 10 light-coloured seeds
(Figure 7.2). In an analogous fashion to carob, the endosperm is
separated from the germ and testa and ground to the commercial guar
gum. The gum powder is available in a wide range of particle sizes. The
coarse particles generally dissolve more slowly in water, a fact which may

Figure 7.3 Tara pods and seeds. The seeds are about 10 mm long and weigh approximately 0.25 g each, of which approximately 18% is galactomannan.

be technologically important, and exhibit lower viscosities. Several manufacturers also offer ranges of thermally degraded gums with reduced viscosities. Steam-treated powders are also commercially available; these exhibit an enhanced rate of solution and a reduction in the otherwise typical bean-like taste.

7.2.3 Tara gum

Although at present of little commercial importance, a gum derived from the tara bush, *Caesalpinia spinosa*, is marketed in Europe (Jud and Lössl, 1986). *Caesalpinia spinosa* is a shrub which is indigenous to Equador and Peru and is grown in Kenya. The reddish fruits are smaller than those of *Ceratonia*, but the seeds are of similar size (Figure 7.3). The gum obtained is similar in viscosity to guar.

7.3 Chemical structure

Galactomannans are a family of linear polysaccharides based on a backbone of $\beta(1\rightarrow4)$-linked D-mannose residues (see Figure 7.4). To this chain, single α-D-galactose residues are linked by C-1 through a glycosidic bond to C-6 of mannose. However, the gum obtained from a single source is not a single substance and it shows wide polydispersity (Doublier and Launey, 1981; Lopes da Silva and Goncalves, 1990), i.e. molecules of differing degrees of polymerisation are present. This is manifest in the difference between the weight average and the number

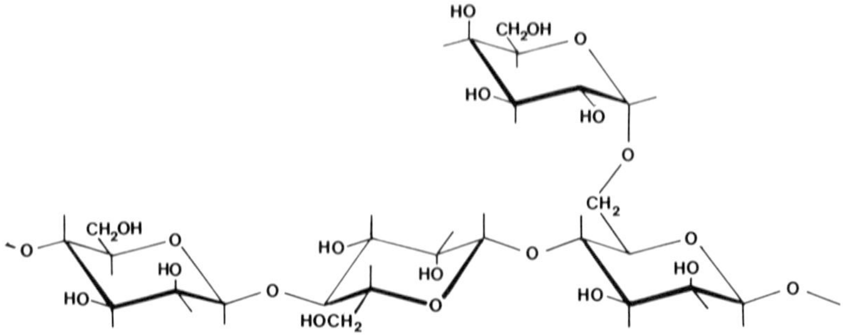

Figure 7.4 Segment of a galactomannan chain.

average molecular weights of galactomannans obtained from these gums (see Table 7.1).

The degree of galactose substitution varies not only from one botanical source to another (for average values see Table 7.1) but also within the molecular species found within one gum. Thus locust bean gum with an average galatose to mannose ratio of 1:4 contains molecules with a lower degree of substitution as well as species which resemble guar with a ratio of 1:2. Such differences will confer on the molecules entirely different properties, and indeed this can be exploited to fractionate the gum (Morris and Ross-Murphy, 1981). The substitution ratio alone does not fully describe the molecules. In such partially substituted molecules there remains the question of the pattern of substitution, and even today this is not clear, although it appears to play a significant role in determining the interaction with other polysaccharides (Dea *et al.*, 1986; Dea, 1990). The majority of the evidence suggests that this is characterised by both ordered and random segments.

7.4 Properties

7.4.1 Solubility

D-Mannose differs from D-glucose only in the stereochemistry at C-2. It is therefore not surprising to find that $\beta(1\rightarrow4)$ D-mannan is an insoluble fibrous material similar to its D-glucose analogue, cellulose. This is believed to be the result of the formation of stable crystalline regions in which the linear chains come to lie in close proximity. The inclusion of side chains spoils this crystallinity and thereby promotes the penetration of water and enhances solubility. Thus, commercial gums showing high substitution ratios such as guar tend to hydrate fully in cold water, while gums with limited substitution, such as locust bean gum, only dissolve

Table 7.1 Comparison of the composition of galactomannans.

Galactomannan source	Average galactose to mannose ratio	Weight average molecular weight ($\bar{M}_w$, daltons)	Number average molecular weight ($\bar{M}_n$, daltons)
Guar gum	1:2	*1.9×10^6	*250 000
Tara gum	1:3	—	—
Locust bean gum	1:4	†1.94×10^6	†80 100

  * From Lopes da Silva and Goncalves (1990).
  † From Hui and Neukom (1964).

Table 7.2 Comparison of the physical properties of galactomannans.

Galactomannan source	Intrinsic viscosity (dl/g)	Viscosity of a 1% solution measured on a Brookfield RVF at 20 r.p.m.		
		Prepared at 25°C, measured at 25°C	Prepared at 85°C, measured at 25°C	Prepared at 85°C, measured at 85°C
Guar gum				
High molecular weight ($\bar{M}_w = 1.9 \times 10^6$)	*14.0	3800	4800	1800
Low molecular weight ($\bar{M}_w = 0.4 \times 10^6$)	*4.5	70	70	20
Tara gum	†11.2	2800	3900	1500
Locust bean gum	‡10.0	200	2000	500

  * From Robinson et al. (1982).
  † From Clark et al. (1986).
  ‡ From Doublier and Launey (1981).

completely in hot water. This behaviour is reflected in the viscosity of the hot- and cold-prepared solutions (see Table 7.2). Galactomannans do not, however, dissolve instantaneously in water. At 25°C guar can take up to 120 min to hydrate fully and yield its maximum potential viscosity. This time can be shortened by using a high-shear mixer or finer powder or by increasing the temperature. However, at temperatures above 80°C, thermal degradation becomes significant and maximum viscosity will not be achieved. The rate and degree of hydration may be dramatically reduced by the presence of other solutes. When using galactomannans in formulated food it is therefore good practice to dissolve these first in any free water present in the recipe before adding the remaining ingredients.

7.4.2 Viscosity

The industrial use of galactomannans is primarily the result of their ability to produce highly viscous aqueous solutions at relatively low concentra-

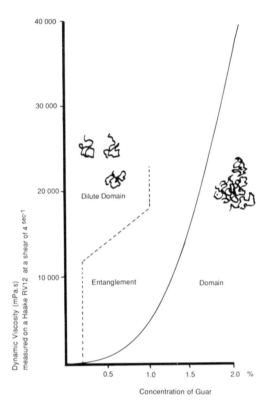

Figure 7.5 Viscosity of guar solutions as a function of concentration.

tions (Goldstein *et al.*, 1973). When galactomannans are dissolved in water, the long mannan chain unfolds to form an open, flexible, non-ordered conformation termed a random coil (Mitchell, 1979; Morris and Ross-Murphy, 1981). At low concentrations these coils are separated by solvent and the viscosity of the solution increases with concentration of polysaccharide in a linear manner, as would be expected by Einstein's theory (Einstein, 1906). As the concentration rises, the statistical chance that the polysaccharide chains will come into contact increases and eventually results in their mutual entanglement. This 'felt' of polysaccharides hinders the mass deformation of the solvent and causes a disproportional increase in the viscosity of the solution. In this concentration range, known as the entanglement domain, the viscosity increases exponentially with concentration (see Figure 7.5). It follows from this model that the longer the molecule and the more extended its conformation in solution, the sooner entanglement will begin and the stronger it will be. In other words, the viscosity will be higher at a given concentration or a lower concentration will be needed for a given viscosity.

The extension of a molecule in solution is related to its hydrodynamic volume, which is the volume in space swept by the molecule when it is allowed to rotate freely about its centre of gravity. This parameter can be measured and is known as the intrinsic viscosity (η), values for which are given in Table 7.2. It will be seen from these values that, as expected for a linear species, the intrinsic viscosity increases markedly with the degree of polymerisation. It follows that galactomannans should be and are available in a wide range of viscosities which reflect the average length of their mannan backbone. However, intrinsic viscosity is also affected by the solvent. Only in thermodynamically good solvents will the polymer chain be optimally extended. In poorer solvents intramolecular binding will be preferred to hydration, and this will limit molecular extension and reduce viscosity. This is particularly important in practical applications where other solutes (sugar, salts or even alcohol) present in the food can reduce solvent quality and influence the rheological properties.

For the efficient use of galactomannans as thickening agents in foods their concentration, C, should lie within the entanglement domain. For galactomannans, this has been found to start when:

$$[\eta]C \approx 1$$

i.e. when the space occupancy or overlap factor is slightly above unity (Doublier and Launay, 1977). Above this concentration theory predicts that the zero shear viscosity, η_0, should rise with the third power of the concentration, so that a doubling of the concentration should result in an eightfold increase in viscosity. With guar it has been shown that the dependency on concentration is far greater, being proportional to the fifth power:

$$\eta_0 \propto C^5$$

and this has been ascribed to association of mannan chains in solution (Robinson *et al.*, 1982). It follows from this observation that the viscosity obtained is extremely sensitive to guar concentration and therefore accurate dosage is essential.

In food development it is important to know not only the viscosity, η_0, of a product standing on the shelf but also if and how this will change when the food is chewed, i.e. it is important to know how the viscosity varies with imposed shear. Where $[\eta]$ is small, i.e. where the polysaccharide molecule is small or compact and

$$[\eta]C \ll 1$$

viscosity is ideally a constant and independent of shear. Translated into food technologist terms this means that by attempting to drink such a liquid at twice the rate it will exhibit twice the resistance. Such behaviour is termed Newtonian flow (see Figure 7.6), and such solutions generally have a slimy and unpleasant mouthfeel.

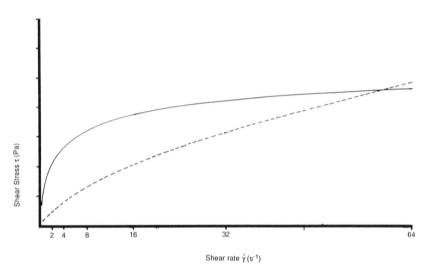

Figure 7.6 Flow curves for guar: —, a 1% solution of a high-viscosity guar, [η] ≈ 14 dl/g, showing marked pseudoplastic flow properties; ---, a 3% solution of a degraded guar, [η] ≈ 5 dl/g, showing predominantly Newtonian flow. Dynamic viscosity $\eta = \tau/\dot{\gamma}$.

With galactomannans and many other polysaccharides which form an extended random coil in solution [η] is large and this does not apply. Such solutions exhibit pseudoplastic flow behaviour (see Figure 7.6) whereby the viscosity falls with shear. In the mouth such thickened solutions appear to thin on mastication and give a pleasant, light mouth-feel. The reason for this shear thinning in the dilute concentration domain is believed to lie in the orientation of the extended molecules under the shear gradient, i.e. they tend to align themselves parallel to the direction of flow. At higher concentrations where the overlap factor is greater than unity, shear has been envisaged to cause disentanglement. As the speed of shear increases with respect to the speed of re-entanglement (or relaxation time of the molecules) so the viscosity, but not the absolute force, falls. Once the shear ceases, re-entanglement is immediate and the solution exhibits its original low shear viscosity. In this respect galactomannans differ fundamentally from starch pastes, which also exhibit pronounced shear thinning but which, once sheared, lose viscosity permanently.

Particularly relevant for the thickening of thermally processed foods is the change in viscosity on heating. Like most viscous fluids, galactomannan solutions reversibly thin on heating, and representative figures are given in Table 7.2.

7.4.3 Interaction with other hydrocolloids

Galactomannans interact with and modify the functionality of many other polysaccharides. Thus hot solutions of xanthan gum and locust bean gum in the ratio of 1:1 form strong elastic gels on cooling; a functionality which neither of the hydrocolloids alone exhibits under normal conditions. The molecular basis of this gelling is believed to be the formation of a three-dimensional polysaccharide network by virtue of the formation of junction zones between xanthan and the naked, or galactose-deficient, segments of the locust bean gum (Cairns et al., 1987). This model is supported by the fact that with increasing galactose content the interaction decreases. Thus tara gum forms only a weak gel and guar produces only a synergistic rise in viscosity. Similar observations can be made with the red algae extracts. Thus, κ-carrageenan, which alone gives a brittle gel, can, by the addition of locust bean gum, be made to form a much stronger elastic gel.

7.4.4 Stability

Depolymerisation of the chain leads to significant reduction in viscosity, and this must be taken into account when using galactomannans in thermally processed foods. Under neutral conditions sterilisation at 120°C for 10 min results in about 10% loss of viscosity. A considerable improvement can be achieved by the addition of trace amounts of sodium sulphite and propylgallate. This suggests that oxidative–reductive depolymerisation is the dominant mechanism (Mitchell et al., 1992). This mechanism does not apply below pH 4.5. At low pH acid-catalysed hydrolysis of the glycoside bond will result in almost 90% loss of viscosity when the solution is heated to 120°C for 10 min. Under such circumstances the use of buffered systems or low temperature processing may be useful. Degradation may also be reduced by incorporating the galactomannans as a coarsely ground powder, thereby delaying hydration.

Galactomannans are usually completely stable to the shear forces encountered in food processing, but under extreme conditions some fission of the mannan chain occurs. Thus high-pressure homogenisation of a 1% guar solution at 200 and 300 bar causes a 5% and 20% reduction in the viscosity respectively.

7.4.5 Interaction with protein

Different macromolecular species, and in particular proteins and polysaccharides, can be thermodynamically incompatible in solution (Tolstoguzov, 1991). Above a critical concentration, typically that at which molecular overlap begins to occur, the two molecular species are no longer totally miscible and the solution separates into two phases, one rich in protein, the other containing the majority of the polysaccharide. Such phase separations are favoured by reducing the solubility of the protein and increasing the intrinsic viscosity of the galactomannan. They occur where gelatin or protein-rich foods (such as milk) are mixed with hydrocolloids and must be taken into account in thickening such systems.

7.5 Applications of galactomannans

The use of galactomannans in formulated foods (Glicksman, 1969) can be directly or indirectly related to three fundamental properties:

1 Their ability to thicken aqueous solutions.
2 Their synergistic interaction with other polysaccharides.
3 Their capacity to control or prevent syneresis.

Of these, thickening ability is the property which is by far the most widely exploited, and for this reason it may be useful to compare the properties of galactomannans with those of starches, which are also widely used as thickeners.

1 Starches are relatively inefficient thickeners and must be used in much higher concentration: 4.0–6.0% as compared with 0.5–1.0% for galactomannans.
2 Starches (with the exception of native high-amylopectic types) produce a very short pasty structure with a full mouthfeel. Galactomannans tend to produce a light, long texture.
3 Much of the viscosity is irreversibly destroyed when starch pastes are sheared. Galactomannans recover fully.
4 Native starches need to be cooked to obtain viscosity. Some galactomannans are cold soluble.
5 The temperature dependence of viscosity is more pronounced for galactomannan solutions than for starches, i.e. better heat penetration in the autoclave.
6 Enzymes which degrade starches are ubiquitous in fresh food. Enzymes which degrade galactomannans are seldom encountered.
7 Starches are fully digested in the alimentary tract. Galactomannans are not degraded and little or no energy is available from their ingestion.

This comparison reveals that in many ways the properties of these two main food thickeners are complementary. It is not surprising, therefore, to find that galactomannans are used with starches to achieve subtle differences in rheological properties which are appreciated by the modern discerning consumer and are the basis of brand differentiation. These rheological effects will also be influenced by other ingredients in the recipe and by the manufacturing technology. Factors such as:

- pH
- Water activity
- Ionic strength

which are recipe dependent, or processing variables such as:

- Shear stress
- Thermal load
- Filling temperature

are all important. Because of this, only rough guidelines for the use of galactomannans can be given, and these will only represent possible starting points for the development of new products. It should also be realised that the problems encountered are such that they can rarely be solved by the use of one hydrocolloid. The optimal solution usually requires the use of a mixture of components with separate functionality and which may undergo beneficial synergistic interactions with each other.

7.5.1 Milk-based products

The use of seed gums in milk-based products is limited by protein phase separation, at high dosages. This is particularly true of acidified or fermented milk products, in which the milk proteins approach or pass through their isoelectric point during the manufacturing process. Other factors which reduce the solubility of the protein will have similar effects. In general, guar is more liable to cause phase separation than locust bean gum by virtue of its higher intrinsic viscosity.

7.5.1.1 Ice cream. The single largest use of galactomannans in milk-based foods is almost certainly in the stabilisation of ice cream (Arbuckle, 1986). Ice cream typically contains between 5 and 15% fat, usually milk fat, 10–12% non-fat milk solids and 14–15% added sugars, the exact composition varying with the fat content. To this mix are added 0.6–1.0% emulsifier–stabiliser mixture which contains up to 0.5% mono/diglyceride and 0.3% high-viscosity locust bean gum or guar–locust bean gum-based mixture. The mono/diglyceride serves to destabilise the protein layer around the emulsified fat globules, which on cooling and whipping at −5 to −8°C leads to a partial churning out of the fat. This free liquid fat, together with fat crystals, forms a stabilising layer around the air pockets (Barfod *et al.*, 1991).

During freezing and whipping about half the water in the aqueous phase freezes, forming fine crystals. During hardening at −30°C these crystals grow until approximately 90% of the water is in the crystalline condition. The smooth texture of ice cream is destroyed if these ice crystals grow too large. This can occur during storage when small temperature fluctuations will result in melting and refreezing of crystals whereby the small crystals will lose mass at the expense of the larger ones. Locust bean gum and guar are known to inhibit the growth of larger crystals, and it has been suggested that this is the result of the binding of the fluid water to the hydrocolloids, which prevents its mass transfer. Owing to its thickening effect on the molten ice cream, it also enhances the creamy mouthfeel of the product. When galactomannans are used as stabilisers, wheying off, i.e. phase separation of the casein, tends to take place on thawing. This can be counteracted by the addition of a small amount (c. 0.02%) of a κ-carrageenan.

Compared with hard pack ice, just described, soft ice is only frozen to c. −5°C and is eaten directly from the freezer. In this case, the need to inhibit ice crystal growth is much reduced and hence, although galactomannans are used to improve mouthfeel in this type of ice, the dosage is usually much lower, typically 0.1%.

7.5.1.2 Fermented milk products. Fresh cheese or quarg is finding increasing use in processed foods. Quarg is obtained by the fermentation of milk or cream, usually with the addition of low levels of rennet. The coagulated casein is cut and separated using traditional cheese-making methods or by centrifugation. When high levels of fat are present the process must be carried out at elevated temperatures (c. 70°C) using special centrifuges to avoid unacceptable loss of fat in the whey. Because this technology is not widely available in the dairy industry, the majority of processed foods use low-fat or fat-free quarg as the starting material. However, when such quarg is subject to shear, as may be necessary in processing, it loses almost all of its structure and reverts to a watery paste. To overcome this problem stabilisers which confer the desired rheological properties on the end product are added. These stabilisers are relatively complex mixtures based on gelatin, carboxymethylcellulose and galactomannan (usually about 0.1%), whose role is to bind the water set free on shear and increase viscosity.

7.5.1.3 Drinks. In comparison with normal milk, milk shakes are thickened to give a creamier mouthfeel, and this also helps to impart temporary stability to the foam by slowing the rise of the air bubbles. Typically less than 0.1% galactomannan is employed in the recipe, which may also include small amounts of xanthan. With chocolate milk, stabilised by weak carrageenan gels, the situation is somewhat different. In this case, the addition of galactomannan tends to destabilise the drink, leading to phase separation. In some cases, addition of a small amount of depolymerised guar is possible, but this will depend on other parameters such as the quality of the milk, cocoa powder, and carrageenan. Generally, galactomannans cannot be used to thicken acidified milk drinks. They may, however, be used in products in which the proportion of milk is low, such as milk-fortified fruit juice drinks. The lower particle density makes stabilisation with pectin or other hydrocolloids difficult, and galactomannans may be added to delay the sedimentation of the casein micelles or agglomerates to an acceptable level.

7.5.2 Desserts

Aerated and non-aerated gelated desserts may be prepared with carrageenans. Galactomannans may be used to modify the texture of the gel and prevent syneresis. In milk-based products c. 0.3% carrageenan and 0.2–0.3% galactomannan are typical addition levels. Slightly higher levels have to be employed in water-based foods. Galactomannans at a concentration of 0.1% may also be added to the cream toppings used on these desserts. This imparts a creamier and fuller mouthfeel, which will be particularly important when the fat content has been reduced.

7.5.3 Mayonnaises

Mayonnaises are oil-in-water emulsions traditionally prepared with 80% oil and egg yolk as the emulsifier. Their characteristic plastic consistency can be attributed to three factors:

1 The small oil droplet size which confers a relative high Laplace pressure and thereby resistance to deformation.
2 The high space occupancy of the oil, which makes contact between the oil droplets obligatory.
3 Attractive forces between the emulsifier layers around the individual oil droplets.

The trend today is to fat- and calorie-reduced products, and here difficulties are encountered in maintaining the desired consistency when the oil content falls below about 60%. In such cases the rheological properties of the continuous aqueous phase must be modified, and this can be achieved by replacing the oil with a starch paste on a *pro rata* basis. The more oil is replaced the more the flow properties deviate from the ideal, and this must be corrected by incorporating a hydrocolloid stabiliser blend in the paste. Typical recipes for retail mayonnaise-like products are as follows:

Soya oil	50.0%	35.0%	15.0%
Starch (chemically modified)	1–2%	2–3%	4–5%
Milk protein emulsifier	1.0%	0.5%	0.5%
Galactomannan	0.4%	0.5%	0.6%
Xanthan	0.1%	0.1%	0.1%

Generally guar with a small addition of locust bean gum is used.

Where the mayonnaise is to be mixed with other ingredients such as fish or meat for the preparation of salads, mayonnaises which will take up fluids from the cut surfaces of these ingredients are required. This can be achieved in wholesale mayonnaises by increasing the galactomannan levels. Salads prepared with fresh vegetables and fruit present a special problem, as these ingredients may release amylases which hydrolyse the starches. For these applications starch-free recipes have been developed based solely on amylase-resistant hydrocolloids. A typical recipe is as follows:

Soya oil	50.0%
Skimmed milk powder	0.8%
10% vinegar	5.0%
Guar, high viscosity	0.5%
Xanthan	0.2%

The appearance of the salad may be improved if the mayonnaise does not drain off the solid salad components. This is particularly true of coleslaw, a fresh salad with a high cabbage content. Such cling effects can be achieved by using a relatively high proportion of xanthan gum to give the guar or locust bean gum-stabilised dressing a flow threshold.

7.5.4 Dressing and ketchups

Ketchups, when prepared from fresh fruit, derive their viscosity from pectin naturally present in the ingredients. Where pectin-deficient or preserved fruit is used, it may be necessary to add starch and/or other polysaccharides. In such products galactomannans can increase viscosity and eliminate syneresis. For example, for hot-prepared, cold-filled (25°C) ketchups gelling blends of xanthan and locust bean gum may be used at dosage levels of approximately 0.4%. These are particularly suitable for products filled in flexible squeeze bottles because they reduce the tendency to drip. Generally, such systems have the disadvantage that temperature fluctuations on storage may promote gelation, however, in this particular application high shear in the nozzle corrects this. In emulsified free-flowing dressings non-gelling mixtures of guar or xanthan and guar are best employed as thickeners. A dressing containing 35% soya oil and 3.5% salted egg yolk could contain 0.4% guar or 0.1% guar plus 0.2% xanthan.

7.5.5 Sterilised soups and sauces

Galactomannans can be used to advantage in the thickening of thermally sterilised soups and sauces. The reduction in viscosity on heating enables heat to be transferred by convection within the container, thus promoting heat penetration and resulting in a reduction in the sterilisation times and thermal load on the product. The small irreversible loss of viscosity owing to thermal degradation of the polysaccharide can be minimised by using a cold preparation and coarsely ground galactomannan, which delays its hydration and degradation. Typically 0.2–0.5% of a blend of xanthan and guar is used.

7.5.6 Deep-frozen foods

The sales of deep-frozen ready meals have increased markedly during the last decade, a trend which is likely to continue with calorie-controlled varieties. These products require that the sauces, dressings and jellies used are freeze–thaw stable. When such aqueous systems freeze, ice formation forces the solutes into a very restricted space between the crystals. In this fluid phase, suspended solid particles will tend to agglomerate and polysaccharide chains will be pressed into close association. Problems arise because these associations or aggregations do not rehydrate or redisperse when the ice thaws. This is manifested as syneresis or separation of serum from the suspension or hydrocolloid-containing phase, which shrinks in size. The addition of a small amount of a cold-soluble galactomannan such as guar can improve freeze–thaw resistance, presumably by sterically hindering the formation of aggregates in the interstitial fluid phase. An example of such a recipe for freeze–thaw-stable cream sauce, hot prepared, is as follows:

Soya oil	60.0%
30% cream	4.0%
Skimmed milk powder	2.0%
Flour	4.0%
Modified starch (adipate)	2.5%
Guar gum	0.5%

7.5.7 Other uses

Galactomannans have found wide use as filling aids for particulate foods, rice and noodle. They are also used in sausage recipes to give the product a better bite and prevent weeping when the frozen product is thawed. In meat tumbling blends, they have been used as a suspending agent for particulate carrageenans. They are used as a binder and phosphate replacer in boneless fish fillet blocks containing farce and can be used to control the viscosity of adhesive batters. Their use in bread and bakery products has been reported. In this case they may serve as a replacement for gluten in dietetic products. Galactomannans have also found use in low-sugar jams, fruit preparations and spreads.

7.6 Conclusion

Galactomannans enjoy wide application in formulated foods. This can be ascribed to their competitive price, and to their wide consumer acceptance. Galactomannans may be described as wholly natural substances which have in no way undergone chemical change or modification. Indeed, as soluble fibre their presence in the diet is known to be beneficial (Lössl, 1989). For these reasons we predict that the use of galactomannans in foods will increase, providing new opportunities for novel products.

Acknowledgements

The author wishes to express his gratitude to Mr Georg Hahn for permission to use some of information contained in this article, and to Mr Paul Ingenpass who placed his knowledge and experience at the author's disposal.

References

Arbuckle, W.S. (1986) Effects of emulsifiers on protein–fat interactions in ice-cream mix during ageing. I. Quantitative analyses. In: *Ice Cream*, Avi Publishing Co., pp. 84–94.

Barfod, N.M., Krog, N., Larsen, G. and Buchheim, W. (1991) *Fat Sci. Technol.*, **93**, 24.

Cairns, P., Miles, M.J., Morris, V.J. and Brownsey, G.J. (1987) X-ray fibre-diffraction studies of synergistic binary polysaccharide gels. *Carbohydrate Res.* **100**, 411.

Clark, A.H., Dea, I.C.M. and McCleary, B.V. (1986) The effect of galactomannan fine structure on their interaction properties. In: *Gums and Stabilisers for the Food Industry*, Vol. 3, G.O. Phillips, D.J. Wedlock and P.A. Williams, eds, Elsevier Applied Science, Amsterdam, pp. 429–440.

Dea, I.C.M. (1990) Structure/function relationships of galactomannans and food grade cellulosics. In: *Gums and Stabilisers for the Food Industry*, Vol. 5, G.O. Phillips, D.J. Wedlock and P.A. Williams, eds, IRL Press, (at Oxford University Press), Oxford, pp. 373–382.

Dea, I.C.M. and Morrison, A. (1975) Chemistry and interactions of seed galactomannans. *Adv. Carbohyd. Chem. Biochem.*, **31**, 241–312.

Dea, I.C.M., Clark, A.H. and McCleary, B.V. (1986) Effect of the molecular fine structure of galactomannans on their interaction properties — the role of unsubstituted sides. *Food Hydrocolloids*, **1**, 129–140.

Doublier, J.L. and Launay, B. (1977) *Industrie Minerale*, **4**, 191.

Doublier, J.L. and Launey, B. (1981) Rheology of galactomannan solutions, *J. Text. Studies*, **12**, 151–172.

Einstein, A. (1906) Eine neue Bestimmung der Moleküldimension. *Annalen der Physik*, 1906, **19**, 289–306.

Glicksman, M. (1969) *Gum Technology in the Food Industry*, Academic Press, New York, p. 130.

Goldstein, A.M., Alter, E.N. and Seaman, J.K. (1973) Guar gums. In: *Industrial Gums*, R.L. Whistler and J.N. BeMiller, eds, Academic Press, New York, pp. 303–321.

Hui P.A. and Neukom, H. (1964) *Some properties of galactomannans. Tappi*, **47(1)**, 39–42.

Jud, B. and Lössl, U. (1986) Tara gum — a thickening agent with perspectives. *Intern. Zeitschrift für Lebensmittel-Technologie und Verfahrenstechnik*, **37(1)**, 28–31.

Lössi, U. (1989) Dietetic aspects of galactomannan. *Ernährung (Nutrition)*, **13**, 10–14.

Lopes da Silva J.A. and Goncalves M.P. (1990) Studies on a purification method for locust bean gum by precipitation with isopropanol. *Food Hydrocolloids*, **4**, 277–287.

Mitchell, J.R. (1979) In: *Polysaccharides in Food*, J.M.V. Blanshard and J.R. Mitchell, eds, Butterworths, London, pp. 51–57.

Mitchell, J.R., Hill, S.E., Jumel, K., Harding, S.E. and Aidoo, M. (1992) The use of anti-oxidants to control viscosity and gel strength loss on heating of galactomannan systems. In: *Gums and Stabilisers for the Food Industry*, Vol. 6, G.O. Phillips, D.J. Wedlock and P.A. Williams, eds, Oxford University Press, Oxford, pp. 303–310.

Morris, E.R. and Ross-Murphy, S.B. (1981) Chain flexibility of polysaccharide and glycoproteins from viscosity measurements. *Techniques in Carbohydrate Metabolism*, **B310**, 1–46.

Robinson, G.R., Ross-Murphy, S.B. and Morris, E.R. (1982) Viscosity–molecular weight relationships, intrinsic chain flexibility and dynamic solution properties of guar. *Carbohydr. Res.*, **107**, 17–32.

Tolstoguzov, V.B. (1991) Functional properties of food proteins and role of protein–polysaccharide interaction. *Food Hydrocolloids*, **4**, 429–468.

8 Modified starches

A. RAPAILLE and J. VANHEMELRIJCK

8.1 Introduction

8.1.1 Sources

There are many types of starch, derived from maize (corn), waxy maize, wheat, potato, tapioca, rice and many other sources. Different starches have different properties and are applied, in the food industry, for nutritional, technological, functional, sensorial and even aesthetic purposes. The thickening and gelling properties of starch have very positive influences on the sensory character of food products and also have important technological/functional effects in industrial processing and in the kitchen preparation of foods.

8.1.2 Availability

Starch is a low-priced and abundant worldwide commodity and an attractive renewable raw material for industry. The most important crops for the production of starch in Europe are maize, wheat and potato.

Maize and wheat are stored dry after harvest and are used by the starch industry throughout the year. In the past most of the maize used by the starch industry was imported from the USA. However, since the beginning of the 1980s, the European Community (EC) has been self-sufficient in cereals, and because the EC imposes a levy on importation practically no maize or wheat is imported any longer.

Potato starch is produced from special potatoes with high starch content, grown mainly in The Netherlands, but also in Germany, Denmark and France. Owing to the high moisture content of potatoes, their transport and storage are problematic. Potato starch is therefore produced in growing areas only, in campaigns which last from September till January. Rice starch is produced from broken rice, which is imported into the EC. Tapioca starch is also imported into the EC, particularly from Asiatic countries.

8.1.3 Quantities

In 1989, a total of 5.1 million tonnes of starch were produced in Europe, of which 59% was maize starch, 21% potato starch and 20% wheat starch. The main part (55%) was used in the food industry, while the remaining 45% found outlet in non-food industrial applications.

8.2 Chemical composition

8.2.1 Structure of starch

The structure of starch and its composition vary according to botanical source. Occurring naturally as small granules — with species-specific grain size and shape — starch is insoluble in cold water. All starches consist of two types of molecules, one known as amylose and the other as amylopectin.

Amylose is a linear-chain molecule consisting of anhydroglucose units connected by α-1,4 linkages (α-1,4-glucan). The degree of polymerisation (DP) ranges from a few hundred up to 10 000, giving molecular weights from 10^4 to 10^6, usually $>10^5$.

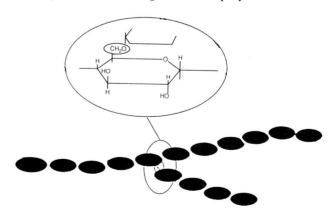

Amylopectin is an α-1,4-glucan with one α-1,6 bond every 20–26 monomer units, giving a branched structure. With a wide variation, the DP is greater than 50 000, yielding molecular weights of approximately 10^7 or more, making it one of the largest natural polymers.

Table 8.1 Main properties of native starches.

Granule	Maize	Waxy maize	Potato	Wheat	Rice	Tapioca
Ratio of amylose to amylopectin	26:74	5:95	22:78	25:75	17:83	17:83
Gelatinisation temperature (°C)	62–74	63–72	56–69	52–64	61–78	52–64
Size (μm)	5–25	5–25	15–100	2–35	3–8	5–35
Starch paste						
Viscosity	Medium	Medium high	Very high	Medium low	Medium low	High
Texture	Short	Long	Long	Short	Short	Long
Clarity	Low	High	High	Low	Low	High
Paste stability						
Deep freeze	Low	Medium	Low	Low	Low	Low
Acid	Low	Low	Very low	Medium	Low	Low
Shear	Medium high	Medium low	Medium high	Medium high	Medium	Medium low

Different enzyme systems are responsible for the synthesis of amylose and amylopectin, and not only the composition but also the ratio of these molecules varies in different plant species (see Table 8.1).

8.2.2 Principal properties of native starches

A knowledge of the properties of the main native starches allows their suitability for a particular application to be predicted. The physical properties of native starches are determined primarily by the ratio of amylose to amylopectin in the starch granule.

The properties of a starch when made into a paste are very important. The consistency of a starch paste correlates with its strength, which can be strong, weak or soft depending on the degree of gelatinisation and the strength of the swollen granules. The texture of a starch paste is determined by the viscoelastic deformation, and can be long, short or stringy. This property depends upon the strength of intergranular linkages and the amount of ruptured granules. The clarity can vary from clear to cloudy to opaque. This property is related to the light-scattering resulting from the amylose association and from other components present in starch.

8.3 Starch manufacturing processes

8.3.1 Basics of separation processes

In plants starch occurs in the presence of other constituents such as protein, fibre, lipids, water and other organic and mineral compounds.

Table 8.2 Approximate composition of cereals/potatoes.

	Maize	Wheat	Potato
Moisture* (%)	7.0–23	14–17	75 (63–87)
Starch* (%)	64–78	57–60	19 (0.8–30)
Protein* (%)	8.0–14	9.0–11	2 (0.7–4.7)
Fibre* (%)	1.8–3.9	2.0–2.5	1.6
Ash* (%)	1.1–3.9	1.5–2.0	1.2 (0.15–2)
Fat* (%)	3.1–5.7	1.8–2.3	0.15

Remainder: hemicellulose (pentosans), carbohydrates
Protein (maize): N × 6.25
Protein (wheat): N × 5.7

*From Whistler *et al.* (1984).

The compositions of some starch sources are given in Table 8.2. Starch occurs in granules, which are different from the other plant constituents in two important aspects: the granules are small, ranging from 2 to $100\,\mu m$, depending on the kind of plant, and their density is high (*c.* $1600\,kg/m^3$). These two properties are used to separate and purify starch in all starch manufacturing processes. The following process steps are applied:

Milling. The raw material is broken up in order to free the starch from the cell in which it is stored.

Fibre removal. The cell walls and outer 'skin' of the raw material are called fibre. Fibre is removed by screening, using the difference in size between the relatively large fibre and the small starch granule. The screening is performed in several stages using intermediate dilution with water in order to wash out as much starch as possible. The washing is done in countercurrent to reduce the volume of water required.

Starch purification. The product that has passed the fibre removal screen consists of starch, other fine particles (mostly protein) and soluble material. Of all these components, starch has the highest density. Starch purification is therefore done in equipment which performs separation on the basis of differences in density. In practice, two types of equipment are used: centrifuges and hydrocyclones. Here also, starch is slurried in water and separated from the other components in several stages. After the last step pure starch is obtained.

Apart from the above three steps in the process, other treatments may be needed to separate specific constituents of the raw material: in the maize starch manufacturing process, after a first coarse beating of the kernel, the germ is separated and maize oil, which has a high nutritional value, is obtained by pressing the germs. In the wheat process, after the (dry) milling, the vital gluten is first removed. This product has a high value as improver for wheat flour used in baking.

The different products obtained by the separation process are in most cases sold in dry form. Solid products first need to be mechanically dewatered and then dried. Soluble material is concentrated by evaporation, and in most cases dried together with the fibre.

All starch processes operate in aqueous media. The water is needed to disperse the different constituents so that they can be liberated. The water also provides the shear forces necessary to obtain a good separation, which would not be obtained if separation was carried out in air. Only starch is produced in extremely pure form; the other products are relatively pure. The wet-milling operation therefore operates within a closed system: fresh water is added for the starch washing, but in all other process steps process water is recycled.

Native starch, although widely used in the food industry, has limited resistance to the physical conditions applied in modern food processing. In order to improve this resistance native starches are chemically modified. These modifications can be achieved in a great number of ways. However, only a limited number are of importance to the food industry, and the major ones are described here. Regular maize, waxy maize, potato and tapioca starch are the primary raw materials used to produce a high number of modified food starches that are tailor-made for individual applications.

8.3.2 Starch modification types and techniques

8.3.2.1 Chemical modification. The modifications usually involve acid hydrolysis, oxidation, esterification or etherification. Multiple treatments may be employed to obtain the desired combination of properties.

Modifications such as cross-bonding, etherification and esterification are conducted in an alkaline slurry reaction at 30–50°C. Reaction times are from 30 min for cross-bonding to 24 h for some etherifications. The product is neutralised in the slurry to a pH of about 5.0, then washed to remove by-products, salts, etc. For taste reasons modified food starches are washed very intensively.

Derivatisation can also be carried out by treating the starch at its natural moisture content with the required reagents and then heating.

Dextrins. Dextrins are manufactured by dry heating (roasting) of native starches at elevated temperatures with or without the addition of acid catalysts. Depending on the process conditions, white or yellow dextrins are obtained.

Acid-thinned starches. Treating a granular starch suspension with an acid results in partially hydrolysed starch molecules, i.e. chains which are shorter than those of the parent starch. This yields starch pastes with low hot viscosities. Initially, acid hydrolysis produces a starch solution with a greater proportion of linear molecules. This in turn will increase the degree of retrogradation since the linear molecules can easily form bundles.

The major food use of these acid-thinned starches is in the field of confectionery, where they are used either alone or in conjunction with other colloids in the manufacture of gums, pastilles and jellies. In these applications, increased set-back leading to the formation of rigid gels gives these starches a significant advantage over native starches.

Oxidised starches. Like acid thinning, described previously, treating starch with oxidising agents also yields starches with chain lengths shorter than those of natural starches. In addition, the hydrogen bonding is affected in such a way as to reduce the tendency to retrogradation, producing soft-bodied gels of high clarity. Oxidised starches are the best thickeners for applications requiring gels of low rigidity. Diluted solutions of highly oxidised starches remain clear on prolonged storage, making them suitable for clear, canned soups.

Substituted starches. These derivatives are mostly produced from waxy maize and sometimes from tapioca starch, because in unmodified form their high amylopectin content provides inherent resistance to retrogradation. In addition, they are easily cooked to give clear, non-congealing gels which are stable over a long period.

However, the native waxy maize and tapioca starch granule has a weak internal structure and is susceptible to thinning at high temperature and/or under acidic conditions with mechanical shear. It is therefore necessary to render the starch more resistant to processing conditions and to reduce the tendency towards retrogradation and syneresis. Substituted starches, suited to specific applications, may be produced from sources other than waxy maize and tapioca.

Cross-linking. The introduction of ester cross-linking groups between starch chains stabilises the granules. The degree of cross-linking is the ratio of the number of bridges to the number of glucose units. The introduction of a very small number of cross-linking groups can modify the starch considerably. In most cases, one cross-link per 500–1000 glucose units is sufficient to achieve the necessary stability without altering the nutritional value.

Esterification and etherification. The introduction of chemical side chains improves starch molecule stability. In other words, chemical modification creates a reduced tendency to gel formation, retrogradation and syneresis when cooking starch-containing foods. In addition, the gel pastes become more transparent.

In esterification and etherification reactions some of the hydrogen atoms of the hydroxyl groups are replaced by a substituent group. As in the case of cross-linking, the degree of substitution is quite low. In many food applications these modifications are combined with cross-linking to create stability during processing conditions and to ensure a long shelf-life.

8.3.2.2 Physical modification.   This type of modification imparts to starch the ability to form a paste in cold water. The process consists of heating a starch slurry, for example on a drum drier, to a temperature above its gelatinisation point followed quickly by drying before re-aggregation or retrogradation of dispersed starch molecules occurs. Pregelatinisation by drum or roll drying is used for both native and chemically modified starches, although most starches are sold in un-gelatinised or granular form. Alternatively, extrusion may be used to produce a wide range of pregelatinised starches. Because of severe process conditions (high shear and temperature) extrusion cannot be applied to products in which high viscosity is required. Roll drying is a much gentler treatment throughout which viscosity remains unaffected.

Heat and moisture treatment involves subjecting a moist starch (e.g. 35% humidity) to heat (e.g. more than 100°C) for several hours. The technique is mainly used with potato starch. The product has an increased gelatinisation temperature and is used for dispersing in boiling water without lump formation.

The search for physical modification techniques that are 'greener' and 'more natural' alternatives to chemical modification continues. Some interesting results have been obtained using starch from specific maize genotypes, but today there is no economically viable alternative to chemical modification to meet the many requirements of food processing in the 1990s and beyond.

8.4 Physical and sensory properties of native and modified starches

8.4.1 Pasting characteristics

A given starch type will yield pastes with characteristic structures and viscosities. The behaviour of starches during cooking can be observed using the Brabender visco-amylograph, an instrument that records the viscosity–temperature profile of a starch suspension during a standard cooking programme. Viscosity changes during gelatinisation and sub-

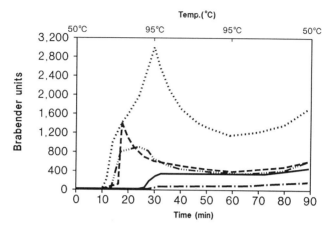

Figure 8.1 Brabender viscosities at pH 5.5, 6% solids, Brabender cartridge 350 cmg: · · · , potato starch; – – –, waxy corn starch; —· · ·—, tapioca starch; —, regular corn starch; —·—, wheat starch.

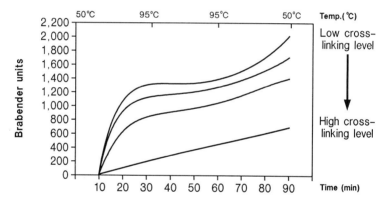

Figure 8.2 Brabender viscosity profiles of acetylated di-starch adipates from waxy maize starch with varying degrees of cross-linking. Concentration, 35 g starch + 450 g distilled water (pH 5.5); cartridge, 350 cmg; cup rotation, 75/min.

sequent cooling are plotted. These Brabender profiles can give an indication of the behaviour of a specific starch under a variety of temperature, pH and shear conditions. Figure 8.1 shows Brabender viscosity profiles for the currently most commonly used native food starches. When starch is modified, with an increasing degree of cross-linking, swelling of the starch granule becomes more difficult. At the same time the peak viscosity decreases. Figure 8.2 shows the Brabender curves of a series of acetylated waxy maize adipates with an increasing level of cross-linking.

8.4.2 Viscosity stability to heat, acid and shear

Viscosity stability during sterilisation can easily be evaluated by measuring the viscosity change of a model, canned 5% starch paste, after autoclaving for 60 min at 121°C. The ratio of the viscosity after processing to the viscosity before processing is an indicator of the stability to heat. The higher the ratio, the better the stability.

Acid stability is determined by simple viscosity measurement before and after acid treatment. The viscosity of a 5% starch paste is defined at neutral pH (7.0) and at pH 2.9 and calculation is made of the ratio of viscosity after acid treatment to viscosity before acid treatment. The higher the ratio, the better the acid stability.

Shear stability is measured by submission of a 4% starch paste to a standard stirring test. Using a high-speed, high-shear mixer such as the laboratory-scale Silverson, the paste is agitated for 5 min at 80°C and 4000 r.p.m. The viscosity measurements and ratio calculation are made as in the previous tests.

One point of great importance regarding all these viscosity tests should be noted. Measurements are made at 20°C, 24 h after the starch pastes have been prepared. This allows time for the starch to swell fully and the pastes to settle. If this time is not allowed, the results of the tests will be inconsistent and unreliable.

Figures 8.3 to 8.8 show the viscosity stabilities of native waxy maize starch and a series of cross-bonded waxy maize starches. The letter code used for di-starch phosphates is analagous to, and in accord with, EC draft directives for modified food starches.

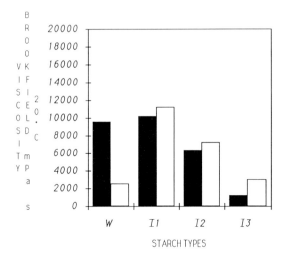

Figure 8.3 Viscosity stability to heat. ■, before treatment; □, after treatment.

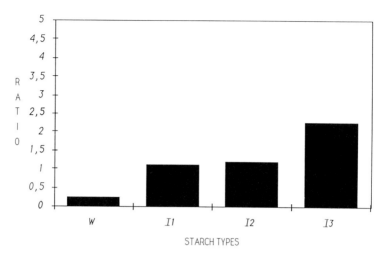

Figure 8.4 Ratio of viscosity after heat treatment to viscosity before heat treatment.

W   Native waxy corn starch
I_1 Di-starch phosphate, low-level cross-bonding
I_2 Di-starch phosphate, medium-level cross-bonding
I_3 Di-starch phosphate, high-level cross-bonding

It must be emphasised that the levels of cross-bonding are 'low', 'medium' and 'high' relative only to each other. The internationally recognised body JECFA, for instance, lists 0.04% phosphorus as the upper limit for phosphate cross-bonded starches.

For a given starch paste concentration heat stability measurements (Figure 8.3) show clearly that, after heat treatment, the absolute viscosity decreases with increasing cross-bonding but the degree of heat stability (Figure 8.4), calculated as the ratio, increases with increasing cross-bonding.

Acid stability results (Figures 8.5 and 8.6) are very similar to those for heat stability, high cross-bonding conferring improved stability, as illustrated by I_3 (di-starch phosphate, high-level cross-bonding).

After high-speed, high-shear mixing (Figures 8.7 and 8.8), viscosity is at its maximum with medium-level cross-bonding, but viscosity stability to shear is optimal with the highest cross-bonding level.

8.4.3 Sensory and textural properties

8.4.3.1 Appearance. Appearance, or visual evaluation, often is the first quality parameter which will impact on anyone making a sensory evalua-

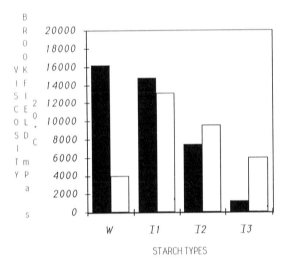

Figure 8.5 Viscosity stability to acid. ■, before treatment; □, after treatment.

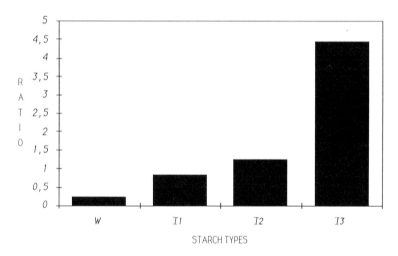

Figure 8.6 Ratio of viscosity after acid treatment to viscosity before acid treatment.

tion of a food product. It is usual to expect that a starch-containing product will be glossy, and whether it is translucent will depend on the type of food under examination, e.g. a salad cream would be glossy and opaque, while a fruit pie filling would be bright and clear. Because of their very low amylose content and, therefore, their almost complete lack of retrogradation, waxy corn starches are more appropriate to the

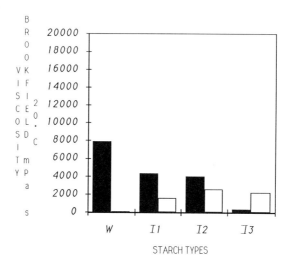

Figure 8.7 Viscosity stability to shear. ■, before treatment; □, after treatment.

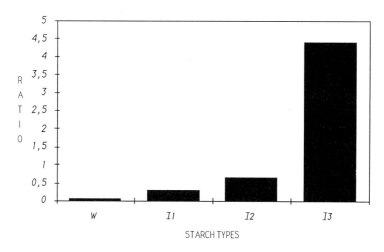

Figure 8.8 Ratio of viscosity after shear treatment to viscosity before shear treatment.

achievement of the required visual effects. Measurement of the light reflection (Figure 8.9) of starch pastes, using a spectrophotometer demonstrates that the opacity of native corn starch pastes increases dramatically with increasing starch concentration. In native waxy and cross-bonded waxy corn starch pastes, clarity is maintained independently of starch concentration.

8.4.3.2 Structure. When spooning a starch-based milk dessert, or stretching or compressing a starch-based confectionery gum, the structure

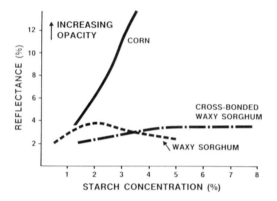

Figure 8.9 Appearance (clarity/opacity) of starches at different concentrations. From Schoch (1900) (See Pomeranz (1985), chapter 2, p. 60).

might be described as short or long, hard or soft, elastic or plastic, etc. The structural properties of any food product are determined by market demands and consumer preferences, how the manufacturer creates the formulation to meet those demands and preferences, and how the supplier can tailor the ingredients to meet the manufacturer's requirements.

Starches are vital ingredients in terms of product structure, and some consideration must be given to the type of starch employed, the kind of modification applied to the starch and the starch concentration used. The choice of starch must be based on functional properties, and the correct choice will have a great influence on the physical properties of the end product.

8.4.3.3 Taste. Taste is highly subjective, highly personal and very much identified with the visual presentation of food products, each of which has its individual taste. The role of starch in taste is passive. Its own taste must be neutral and it must not, in any way, mask the taste associated with the product in which the starch is incorporated. In all cases, consideration of the formulation is vital in deciding the starch type and concentration to be employed.

8.4.3.4 Texture. Over the last 20 years, significant progress has been made in the physical determination of the textural parameters of food products. The Instron dynamometer, with some adaptation, is a very good tool for the determination of a texture profile.

A food product is subjected to two successive compressions to derive a force–time diagram (Figure 8.10), from which a series of textural parameters can be derived, namely hardness, cohesiveness, gumminess, springiness, chewiness and adhesiveness. Using these parameters, results

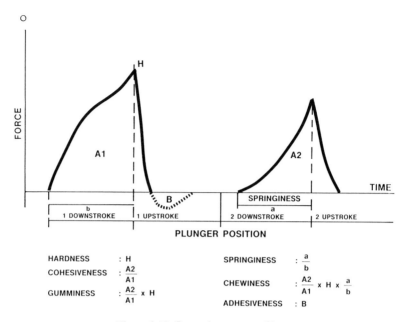

Figure 8.10 General texture profile.

Table 8.3 Texture of starch pastes.

Starch			General texture profile of starch paste (5% dry solids, determined after 20 h)				
Code	Modification type	Level	Hardness (G)	Cohesiveness	Gumminess (G)	Chewiness (G)	Springiness
W	—	—	16	0.9	14	13	0.9
L_1	Phosphate	Low	42	0.74	31	26	0.82
L_2	Phosphate	Medium	26	0.76	20	17	0.85
H_1	Adipate	Low	31	0.6	18	15	0.8
H_2	Adipate	Medium	20	0.7	14	11	0.8

from physical testing, with the Instron, have been found to correlate very well with results from taste panel evaluations based on the same parameters.

Some important texture evaluations (Table 8.3) have been made on model systems of 5% starch pastes of native waxy corn starch and a series of modified waxy corn starches. Evaluation commenced 20 h after the pastes had been prepared. The results showed that hardness increases significantly with increasing cross-bonding and that cohesiveness de-

Table 8.4 Effect of cooking on starch pastes.

	Appearance	Texture	Stability	Viscosity
Undercooked	Cloudy	Thin starch	Poor	Low
Optimal cook	Clear	Heavy-bodied Short texture	Good	Good
Overcooked	Clear	Loss of cohesion Long-textured	Fair	Viscosity loss

Critical parameters
  Temperature–time relationship
  Concentration/shear/pH
  Level of cross-bonding

creases with increasing cross-bonding. It should be noted that acetylated waxy di-starch adipate gives a softer, less cohesive product which, consequently, looks and feels creamier.

There is great variation, and great potential for variation, in textural parameters. The choice of an appropriate starch, in the proper formulation and at the correct concentration, can greatly influence and alter the texture of food products.

8.5 Preparation and use

8.5.1 Optimal preparation conditions for starch pastes

Unmodified and modified cook-up starches require heat to cause the starch granules to hydrate and swell. The degree of hydration varies with time, temperature, acidity and shearing action. The more hydrated a starch granule is, the greater will be the viscosity, clarity of the solution and the degree of retrogradation upon cooling.

8.5.1.1 Cooking of starch pastes: temperature–time relationship. The objective in cooking a starch paste is to achieve optimum appearance, texture, viscosity and stability. Undercooking and overcooking lead to undesirable effects, as shown in Table 8.4. Clearly the levels of cross-bonding on the starch must be matched to the temperature–time profile.

8.5.1.2 Shear rates. The processing conditions will dictate the level of cross-bonding required. In some salad dressing products, e.g. relishes, prepared in steam-jacketed vessels there is a low shear factor and a lightly modified starch can be used. In ultraheat-treated (UHT) custard produced by direct steam injection there is a high shear factor. A highly modified starch is thus required. In canning, a medium shear factor is found when using an agitating or a hydrostatic retort.

8.5.2 Influence of other food components

8.5.2.1 Acidity. Acids disrupt the starch granule and aid gelatinisation. This is important, and highly cross-bonded starches are required in, for example, certain fruit systems. In milk and meat products there is no 'pH assistance', which means that lower cross-bonded starches can be used unless, as in UHT products, there is a high shear factor.

As described earlier, the pH is a critical factor in canned foods.

8.5.2.2 Sugars/soluble solids. When high levels of sugar/soluble solids are used in conjunction with starches (e.g. fruit fillings) there is competition for the available water. The sugars will preferentially take up the water, leaving insufficient for full starch gelatinisation. The solution is to hold back part of the sugar until the end of the cook cycle.

8.5.2.3 Fats/proteins. Fats and proteins will coat the starch granule, thereby inhibiting full gelatinisation. The order of ingredient addition is thus critical.

8.6 Major commercial applications

The main reasons for using starch in manufactured foods are as follows: to provide thickening; to influence the texture; to influence the appearance and to act as a filler. For assessing the suitability of starches for this purpose, it is convenient to consider manufactured foods in two broad categories: powdered products, i.e. dry mixes, and processed products, i.e. liquid, gel or paste forms.

8.6.1 Confectionery

Starches used in the confectionery industry have two different functional roles. In their gelled state, either alone or in conjunction with other hydrocolloids, they form the basic structure of gums, pastilles and jellies. In the granular state they are used in paste goods, for dusting purposes or for moulding.

Modified starches have excellent thickening and gelling properties and are used in the production of a large range of gums. The properties these starches impart to the gums depend on the source of the base starch and the type and level of modification. The main base starches are

Table 8.5 Composition and properties of starch-based gums and jellies.

	Starch	Starch/gelatin	Starch/gum arabic	Starch/pectin
Starch				
Types	Thin-boiling Oxidised Acetylated Cold-soluble	Thin-boiling Cold-soluble	Oxidised	Cold-soluble
Dose (%)	10–30	2–12	*c.* 20	1–1.5
Other gelling agents				
Types		Gelatin 150–180 Bloom	Gum arabic	Pectin
Dose (%)		3–4	*c.* 20	1–1.5
End product				
Humidity (%)	8–13	12–15	8 to 9	20–22
Structure	Soft to hard	Soft	Hard	Soft
Texture	Short to long	Medium long	Long	Short
Clarity	Medium to high	Medium	Clear	Clear

maize, waxy maize and potato. The most popular modification for confectionery applications is acid thinning, particularly when applied to maize starch. In this way, thin-boiling starches of different viscosity levels can be obtained depending on the degree of acid hydrolysis. Oxidation is another way to reduce starch viscosity: oxidised waxy maize starches give hard and clear gums and are currently used in combination with gum arabic. Acetylation is sometimes applied to slow down starch retrogradation. Acetylated starches are used in the production of salted liquorice gums. Cold-soluble starches are the most recent generation of starches used in gum confectionery.

All of the above mentioned starches can be used alone or in combination with other hydrocolloids. Table 8.5 gives an overview of the usage levels of the different starches and other gelling agents and also the final properties of the gums and jellies.

Wine gums are obtained by combining the boiling starch with gelatin. Oxidised starch can be combined with gum arabic and cold-soluble starch with pectin. In addition to modified starches and/or other gelling agents, the other main ingredients used in the production of gums and jellies are sugar and glucose syrups.

In the presence of water, starch-based gum compositions are concentrated by boiling to a dry substance of 68–72%. The traditional method of cooking is at atmospheric pressure using open steam-heated pans. Current cooking methods are continuous, heating the premix by direct steam injection (jet cooker) or indirect heating (heat exchanger). After cooking, the gum masses are deposited in mould impressions in

Table 8.6 Hard and soft starch gum formulations for jet cooker.

	Hard gum	Soft gum
Sucrose (%)	25	32
Glucose syrup (%)	25	32
Acid-thinned maize starch	—	12
Oxidised waxy maize starch	30	—
Water (%)	20	24
Theoretical dry solids (%)	71	68
Premix viscosity at 70°C (mPa s)	<500	<300

Table 8.7 Canned foods 1.

Application	Starch type	Properties/purpose
Soups Sauces Gravies	(Acetylated) di-starch phosphates	Viscosity stability against heat/shear/acid
	Acetylated di-starch adipates	Provides texture/gloss/transparency Imparts mouthfeel
	Pregelatinised native waxy maize starch	Mechanical dosing of particulates
	Oxidised waxy maize starch	Body/transparency/no viscosity
Baked beans Pasta	Di-starch phosphates	Viscosity stability against: heat/shear Provides texture Imparts mouthfeel

dried moulding starch. The gums are dried in stoves to the desired moisture level, demoulded, and finally surface treated (with sugar sanding or pan coating) before packing. Table 8.6 gives formulations of hard and soft starch-based gums.

There is a current trend in gum manufacturing to increase the dry substance of the gum mass before moulding in order to reduce the stoving time as much as possible. For some gum types the cooked gum masses can reach 78% dry substance. In these cases, gum compositions and gelling systems need to give sufficiently low moulding viscosities but, after drying, should give the right texture, clarity and shelf-life. Adding specific cold-soluble starch derivatives can contribute significantly to the achievement of these objectives. These products are used in combination with classical confectionery starches or with other gelling agents such as gelatin and pectin.

Recent developments in cooking and depositing technology also help to achieve these goals. Tubular cookers have become more efficient, assuring an accurate and constant control of the temperature of cooking and gelatinisation of starch. Specifically designed depositing systems have been developed, allowing higher moulding viscosities and higher moulding temperatures. Extrusion cooking is another process used to cook gum masses to very high dry substance concentrations. Cooking gum masses to the final dry substance concentration while retaining depositing viscosities is the aim of all confectioners. Modified high-amylose starches offer possibilities. However, with these starches, gums become completely opaque, which is very often considered an undesirable property.

Development work in the starch industry is continuing in order to obtain starch-based high dry substance gelling systems with good clarity and acceptable textures.

8.6.2 Canned and bottled foods

This group comprises a wide range of food types including vegetables, pet foods, fish, meats, fruits, soups, baked beans, baby foods, pasta, sauces, desserts and pie fillings.

8.6.2.1 Canning: technical principles. The principle of canning is to pack and preserve a food in prime condition so it can be stored under normal ambient conditions for long periods (up to several years). In order to achieve microbiological stability, the sealed cans are submitted to a thermal process that destroys or renders inactive all micro-organisms that can cause toxicity (food poisoning) or spoilage. The cans are then cooled and dried. The *pH of the food* is critical in determining the extent of the heating process.

8.6.2.2 Properties of starches for use in canned and aseptic filled foods.

1 *Water binding.* The starch must have adequate water-binding properties to minimise syneresis.
2 *Thickening.* The characteristics/product image dictate the amount of thickening required. In addition to viscosity, texture and mouthfeel are critical.
3 *Heat penetration.* Heat penetration rates vary between the unmodified starch types (tapioca and waxy maize are the fastest, maize the slowest). With modified starches, the higher cross-bonded starches allow faster heat penetration.
4 *Milk protein protection/compatibility.* With low levels of milk/cream (e.g. creamed soups), the starch-thickened system can protect the protein, giving a smooth texture. In custards, however, hydrolysis of the acetyl group in acetylated starches leads to reaction with the proteins, resulting in granularity. Hydroxypropyl starches overcome this problem.
5 *Stability during sterilisation.* The correct choice of starch will ensure

Table 8.8 Canned foods 2.

Application	Starch type	Properties/purpose
Hot meats	Di-starch phosphates Acetylated di-starch adipates	Binding and thickening agent Texture and mouthfeel

Pie fillings	Acetylated di-starch phosphates
	Acetylated di-starch adipates
	Hydroxypropylated di-starch
	  phosphates
		Viscosity stability against heat/acid
		Provides texture/gloss/transparency
		Good to excellent freeze–thaw
		  stability

UHT custards
  and puddings
	Hydroxypropylated di-starch
	  phosphates
		Thickening agent
		Good shelf-life
		Creamy body
		Gloss

Table 8.9 Bottled foods 1.

Application	Starch type	Properties/purpose

Salad dressings,
* salad creams,*
* mayonnaise*
Hot preparation
	Native maize starches and/or
	  acetylated di-starch adipates
	Di-starch phosphates
		Viscosity stability against heat/
		  shear/acid
		Influence on texture: creamy/
		  short/flowability
		Mouthfeel and gloss

Cold preparation	Pregelatinised native starches
	  and/or pregelatinised acetylated
	  di-starch adipates
	Di-starch phosphates
		Viscosity stability against shear
		  and acid
		Shelf-life
		Influence on texture: creamy/
		  short/flowability

Table 8.10 Bottled foods 2.

Application	Starch type	Properties/purpose

Sauces, gravies
* ketchup*
Hot preparation
	Di-starch phosphates
	Acetylated di-starch
	  adipates
		Viscosity stability against heat/shear/
		  acid
		Thickening
		Stabilising

Cold preparation	Pregelatinised di-starch
	  phosphates
	Acetylated di-starch
	  adipates
		Texture: creamy/short/pulpy/flowable
		Mouthfeel

Table 8.11 Bottled foods 3.

Application	Starch type	Properties/purpose
Pickles	Di-starch phosphates	Thickening
Relishes	Acetylated di-starch adipates	Suspension of particulates
Baby foods	Acetylated di-starch adipates	Viscosity stability against heat/ shear/acid
		Excellent freeze–thaw stability
		Long shelf-life

viscosity stability under the heat and shear conditions found in the can.

6 *Shelf-life stability*. In addition to microbiological stability the physical and textural properties of the product must not adversely change during storage. Correct starch usage can ensure this does not happen.

7 For bottled foods, the choice of starch is also governed by its ability to withstand the acidity of the system and the shear during either hot or cold processing.

A multiplicity of starches are used. These range from unmodified maize and waxy maize to all classes of modified starches. A summary of the use of these starches in canned and bottled foods is given in Tables 8.7 to 8.11.

8.6.3 Frozen foods

A significant sector of the food industry today involves product storage under frozen or chilled conditions. In products which contain starch-thickened sauces or gravy, an additional demand is made upon the functionality of the starch.

Retention of starch gel systems under cold storage hastens the retrogradation mechanism, with the concomitant dangers of syneresis and destabilisation. Although systems based on waxy maize starch (amylopectin) are superior to amylose-containing starches, such as wheatflour and cornflour, they are still subject to retrogradation on cold storage. In its mildest form this could be manifest as the development of a cloudy appearance in a fruit pie filling, where good clarity is sought.

A common method of avoiding such problems is to superimpose a further modification in addition to the cross-linking. This second reaction may be an esterification, for example, the introduction of acetyl groups on to the starch molecule, or etherification by reacting it with propylene oxide. It would appear that the insertion of substituent groups on the starch chain effectively interferes with the aggregation of molecules which normally causes retrogradation.

Table 8.12 Frozen foods.

Application	Starch type	Properties/purpose
Pie fillings	Acetylated di-starch adipates Hydroxypropylated di-starch phosphates	Viscosity stability against heat/ shear/acid Freeze–thaw stability Texture Clarity
Desserts	Hydroxypropylated di-starch phosphates	Stabilising Thickening Improved shelf-life
Battered products	Di-starch phosphates Acetylated di-starch phosphates	Uniform coating Good cohesion Good adhesion

Frozen products place special demands on starches: they must not only withstand processing (heat, shear, acid conditions) but also retain the desired paste or gel characteristics (clarity, absence of retrogradation) during the storage life of the food product. Table 8.12 shows the properties of starches required for application in frozen food products.

8.6.4 Dairy desserts

These milk-based foods are produced in many forms with a range of consistencies, from firm-textured products through thick, spoonable products to thin, pourable toppings. Water binding and thickening are the main functional properties required of all types of milk desserts. The starch must have adequate water-binding properties to minimise syneresis. The characteristics of the end product dictate the amount of thickening required. In addition to viscosity, texture and mouthfeel are critical. A whole range of starch-thickened milk desserts is available to the consumer.

8.6.4.1 Domestic preparations. The earliest attempts at convenience puddings were the dry mix, cook-up starch puddings for domestic preparation. The classical starch thickening and gelling agent is maize starch, although combinations of maize and other starches are possible, and products based on other starches such as wheat, arrowroot, rice, etc. are known. The maize starch product is simple but versatile and can be eaten as a mouldable pudding or a thick pouring sauce depending on the concentration used. Both the pudding and the sauce can be eaten cold or warm. This product can also be used as a cake filling. Two attributes are probably responsible for this dessert's popularity: its good eating quality and its potentially low cost. The lowest cost products contain only starch, colour and flavour, and are sold in simple packages. The cook, in this case, has to add both sugar and milk.

Puddings of this type have a short, but heavy, almost sticky texture, and the use of milk in their preparation gives them a good nutritional image. They can be unmoulded, but with some difficulty. The products thicken rapidly on cooling but a relatively long time is needed before true gelling occurs as a result of retrogradation of the starch amylose fraction. At this point products can be removed from the mould.

When starch is combined with carrageenan, desserts become less pudding-like and have more flow-type properties.

Two alternative preparation methods are normally recommended for puddings and flan-type desserts made from mixes to which sugar must be added. The powder and sugar are blended together until homogeneous, then a little cold milk is added and the mixture is blended to a smooth paste. The remainder of the milk is brought to the boil and either:

- The hot milk is added to the paste, blended thoroughly, then poured back into the pan used to heat the milk. The complete blend is then returned to the boil and simmered for 1–2 min

or

- The smooth paste is added to the freshly boiled milk carefully with stirring and simmered for 1–2 min.

8.6.4.2 Ready-to-eat desserts. Starches and carrageenan have found wide application in the area of ready-to-eat milk desserts; they are inevitably the hydrocolloids of choice for these products. Therefore, after describing the various processes used to produce milk desserts on the industrial scale, the effect of the applied heat, mechanical and filling treatments on the above-mentioned texturing agents will be discussed and also the extent to which they affect the texture of the final product.

The process of manufacturing milk desserts involves heat and mechanical treatments. Various heat treatments can be applied: pasteurisation or sterilisation by either rotary retorting or an ultrahigh-temperature short-time (UHTST) process. The objective of such heat treatments, apart from ensuring food safety, is to extend the shelf-life of the milk desserts by reduction/destruction of micro-organisms and by inactivation of enzymes, at the same time preserving as much as possible the sensory properties and nutritional value.

In addition to water binding and thickening, starches control the following properties in ready-to-eat desserts:

Heat penetration. Heat penetration rates vary between the unmodified starch types (tapioca and waxy maize are the fastest, maize the slowest).

With modified starches, the higher cross-bonded starches allow faster heat penetration.

Stability during sterilisation. The correct choice of starch will ensure viscosity stability under the heat and shear conditions found in the can.

Shelf-life stability. In addition to microbiological stability the physical and textural properties of the product must not change during storage. Correct starch usage can ensure stable properties on storage.

As UHTST-processed milk desserts are becoming the most common type, more details will be given on the use of the starch in these products. As already discussed, UHTST processing is basically the processing and sterilising of liquid products at high temperatures (e.g. 140°C) for very short times (3–10 s). The utilisation of UHTST and aseptic packaging (packaging into sterile containers under sterile conditions) is known as aseptic processing. It has many advantages over classical heat-processing techniques.

The major advantages of aseptic processing are long product shelf-life and stability at ambient temperatures, lower energy costs for processing and improved organoleptic properties (taste, flavour) compared with canned products discussed earlier. Lighter weight, less expensive containers and lower shipping and storage costs are further benefits.

The various types of heat exchangers that are now available for continuous processing with short time application of ultrahigh temperatures are: plate heat exchangers, tubular heat exchangers, scraped-surface heat exchangers. Besides these indirect heating systems, direct steam injection into the liquid food product is also used.

As mentioned previously, two hydrocolloids are currently used in ready-to-serve desserts: starch and carrageenan. Starch is used as a thickening agent; it provides 'body' and full mouthfeel to the dessert. Carrageenan provides a wide variety of textures depending on the type and concentration of the carrageenan in combination with the other dessert ingredients. A typical composition of a vanilla-flavoured dessert is given in Table 8.13.

Table 8.13 Typical composition of a vanilla-flavoured dessert.

Ingredients	Weight (g)
Milk (3.1% fat)	81.5–83.0
Sucrose	8.0–12.0
Skimmed milk powder	1.8
Starch	1.5–4.5
Carrageenan	0.15–0.25
Vanilla flavour	0.025
Colour	0.005

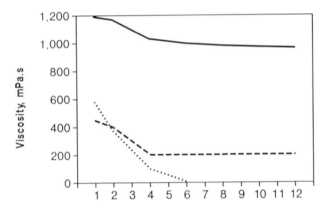

Figure 8.11 Evolution of viscosity during 12 weeks' storage at 20°C. · · · native starch; – – – phosphated waxy starch; — acetylated waxy adipate.

The individual process steps in UHTST dessert manufacture and the process parameters after the thickening/gelling affect the texture of the final product. An example of a typical manufacturing procedure is as follows:

1 After blending all the dry ingredients, add a small quantity of milk (at 4–7°C) to produce a homogeneous slurry. Then add this premix to the rest of the cold milk and mix thoroughly until complete dispersion of the ingredients is obtained. For this purpose, no heating is required.
2 Preheat the mix to between 70 and 90°C and hold at that temperature for 10 min.
3 Homogenise (before or after UHTST treatment). No holding time.
4 Hold the mix at preheating temperatures.
5 Apply UHTST treatment, e.g. at 142°C for 5 s.
6 Cool to filling temperature and store at filling temperature.
7 Fill aseptically either hot (70°C) or cold (<10°C), e.g. in polyethylene bags, tetrabrick, etc.
8 Store in ambient conditions.

In order to obtain the optimum texture and mouthfeel from the starch, its extent of gelatinisation must be well controlled. The extent to which starch is heated during the preheating step and afterwards during UHTST treatment will affect its degree of gelatinisation. When starch is optimally swollen by the end of the UHTST process, the optimum texture will be found in the final product.

The choice of starch type is very important to assure good stability of the prepared UHTST desserts on storage. The graphs in Figures 8.11 and

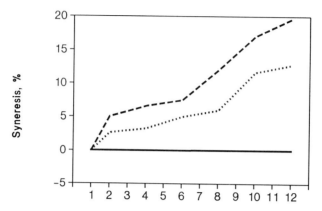

Figure 8.12 Evolution of syneresis during 12 weeks' storage at 20°C. · · · native starch; − − − phosphated waxy starch; — acetylated waxy adipate.

8.12 show the evolution of the viscosity and syneresis during a storage period of 12 weeks for creamy desserts prepared by UHTST indirect heating. It is obvious that the acetylated waxy adipate shows better behaviour (i.e. good viscosity stability and no syneresis during storage) than the phosphated waxy starch and the native starch blend.

Studies of skimmed milk-based UHTST liquid puddings, using an Alfa-Laval VTIS pilot plant, have shown that desserts which contain a higher starch (4.5%) and lower carrageenan level (<0.1%) and which are sterilised by direct heating under severe conditions (142°C, 15 s) may more easily foul the equipment. Although fouling is a reflection of increased steam pressure, it is also related to product formulation. Hydroxypropylated waxy maize phosphate has a superior performance, i.e. causes less fouling, than acetylated waxy maize starch adipate under the same conditions. However, studies have shown that it is necessary to have a minimum level of 3% starch in a formulation to obtain sufficient body and mouthfeel in the dessert.

Table 8.14 summarises the optimal starch types and process conditions for UHTST-sterilised dairy desserts.

8.6.5 *Food powders*

These range from the traditional custard and blancmange powder, dry soup and sauce mixes to the more recent adjunct mixes for stews and casseroles. Usually, the starch is present in its granular form and, hence,

Table 8.14 Summary of optimal conditions for different UHT systems.

Ingredients and process parameters	Direct steam heating	Indirect heating (plates)	Swept surface heating
Starch type and concentration			
Gelled desserts	3% Acetylated di-starch waxy adipate 3% Di-starch phosphate	3% Acetylated di-starch waxy adipate	3% Acetylated di-starch waxy adipate 3% Di-starch waxy phosphate
Creamy desserts	3%Acetylated di-starch waxy adipate	3% Acetylated di-starch waxy adipate	3% Acetylated di-starch waxy adipate
Preheating		65–76°C, no holding	
Homogenisation	Not recommended to maintain texture and mouthfeel	Before UHT Treatment c. 20 kg/cm² to avoid sandiness	Before UHT Treatment c. 75 kg/cm² to avoid sandiness
Heat shock	Direct steam 142°C/5 s	Indirect 140°C/10 s	Indirect 138°C/5 s

poses no stability problems during storage. Occasionally, a reduced starch moisture content is desirable if other sensitive ingredients are included. These food products are reconstituted just prior to consumption using mild conditions of heat and shear, and therefore, even in the swollen state, the starch presents no serious technical problems. For this reason unmodified native starches are widely used in the formulation of these powdered, dry mix products. There is a group of products within this category in which reconstitution and thickening is carried out by a cold process. Typical of these is the 'instant dessert' prepared by the addition of cold milk or water.

In these circumstances it is necessary to use a 'pregelatinised' or 'instant' starch. As already described, pregelatinisation is a physical modification in which the starch slurry is passed between heated rollers. The process has the effect of gelatinising and drying the starch simultaneously. In this way the starch is capable of reconstitution and thickening without further heating. As previously stated, the desired functionality of starch does not always involve thickening. Sometimes the opposite is needed, bulk and volume in the powder mix without undue thickening power. In this case the starch acts as a filler and inert bulking agent. In this role it frequently serves an additional function of aiding dispersion of the powdered mix upon reconstitution. Acid-thinned starches and oxidised starches are used for this purpose. Table 8.15 summarises the

Table 8.15 Dry mix foods.

Application	Starch type	Properties/purpose
Instant puddings and desserts	Pregelatinised di-starch phosphates	Instant thickening Stiff or creamy texture Glossy appearance
Cook-up puddings	Native maize starch	Stiff texture
Soup and sauce mixes (instant)	Native maize starch Oxidised waxy maize starch Di-starch phosphates (pregelatinised)	Body Mouthfeel Thickening
Bakery mixes	Native starches (pregelatinised) and (acetylated) di-starch adipates	Texture Shelf-life Binding Thickening
	Granular starches	
Baking powder Pudding powder Biscuits	Regular maize starch	
Baking powder Cakes Biscuits	Wheat starch	
Pastry Pudding powder	Waxy maize starch	
Bakery jams Pie fillings	Cross-bonded acetylated waxy starch	
	Pregelled starches	
Bakery mixes Choux pastry	Roll-dried maize starch	
Almond dough Choux pastry	Roll-dried waxy maize starch	
Instant puddings Instant pie fillings	Roll-dried cross-bonded waxy maize starch Roll-dried cross-bonded acetylated waxy maize starch	
Instant puddings	Roll-dried cross-bonded acetylated potato starch	

optimal starch types and their functionality in a range of powdered food products.

8.6.6 Bakery products

Several types of starches are used by the bakery and biscuit industry (Table 8.16). Native starches from maize and wheat are used in baking

Table 8.16 Bakery products.

Starch type	Application
Granular starches	
Regular maize starch	Baking powder, pudding powder, biscuits
Wheat starch	Baking powder, cakes, biscuits
Waxy maize starch	Pastry, pudding powder
Cross-bonded acetylated waxy starch	Bakery jams, pie fillings
Pregelled starches	
Roll-dried maize starch	Bakery mixes, choux pastry
Roll-dried waxy maize starch	Almond dough, choux pastry
Roll-dried cross-bonded waxy maize starch	Instant puddings, instant pie fillings
Roll-dried cross-bonded acetylated waxy starch	
Roll dried cross-bonded acetylated potato starch	Instant puddings

powders to prevent premature reaction of the acid and the basic chemicals. The dose used varies between 25 and 55%. Regular maize starch is also an important ingredient of cooked bakery puddings. Sometimes waxy maize starch is blended in. In dough products, maize and wheat starch contribute to the texture of biscuits. They become less hard with improved bottom and less cracking. There is also less tendency for browning. The above effects are found in different types of biscuits, either wire cut, rotary moulded or deposited.

Wheat starch contributes to the quality of cakes. Partial substitution of wheatflour by wheat starch gives a lighter and constant cake volume, with improved texture. Cross-bonded and acetylated waxy maize starches, particularly waxy maize adipates, are currently used in bakery jams and pie fillings. These products should have a dry substance of more than 65%. Such starches contribute significantly to the bake-stability as well as to the freeze–thaw stability of the jams and fillings. In some formulations bake-stable pectin is added.

Pregelled maize starch can be used in instant bakery puddings, however when specific properties are needed roll-dried waxy starch adipates are recommended (see later). Roll-dried maize starch is a main ingredient of instant choux pastry mixes. Sometimes roll-dried waxy corn starch is also used. In recipes containing a high percentage of oil it is necessary to have a more coarsely milled pregelatinised starch to obtain a better 'fat binding'. These starches can also be used in almond dough to prevent drying out of the paste. Roll-dried cross-bonded waxy maize starch and roll-dried cross-bonded acetylated waxy maize starch are currently used in instant bakery puddings and instant pie fillings. In bakery puddings roll-dried cross-bonded acetylated potato starch is also used. These starches are often used in combination with sodium alginate, or with carrageenan. A typical instant bakery pudding formulation is given below:

Ingredients

Roll-dried waxy maize adipate	100 g
Icing sugar	212 g
Whole-milk solids	80 g
Sodium alginate	7 g
Colour, flavour	1 g
	400 g

Method

1 Mix the ingredients.
2 Add the ingredients blend to 1 litre of cold water while stirring.
3 Mix at medium and high speed until the setting is sufficient.

The property requirements of these products are:

- Good depositing directly after mixing
- Direct setting
- Good cuttability
- Bake-stability
- Freeze–thaw stability
- Storage stability (no water separation after 48 h)
- Short texture
- Fine structure.

Instant pie fillings can be prepared with roll-dried di-starch phosphate or acetylated di-starch adipate. Depending on the modification, the pie filling is more or less glossy with a long or a short texture. A di-starch phosphate gives a different texture and more gloss than an acetylated waxy adipate.

8.6.7 Miscellaneous

In addition to the applications which have been listed above there are a number of minor but nevertheless important food applications for starches which should not be overlooked such as:

- Batter starches to improve adhesion and breading pick-up
- Starches in mixes for meat processing to provide the required texture, water binding and consistency
- In glazes starch is used for its film-forming properties and to achieve an attractive sheen
- In snacks (extruded) starches perform as binder, texturiser and provide a good expansion

- Lipophilic starches can be applied in emulsified food products for their emulsifying and fat-binding properties
- Agglomerated instant starches for microwaveable foods.

Acknowledgements

The authors gratefully acknowledge information provided by Mr P. Lawson of Cerestar UK, Manchester, Mr M. Fitton and Mr F. van Esch of Cerestar R&D, Vilvoorde (Belgium) and Mr V. De Coninck and Mr H. Timmerman of Cerestar Eurocentre Food, Vilvoorde (Belgium).

Bibliography

Fawcet, P. (1985) Starches purely functional. *Food*, January, pp. 20, 21, 26.

Light, J.M. (1990) Modified food starches: why, what, where and how? *Cereal Food World*, **35(11)**, 1081–1092.

Morris, V.J. (1990) Starch gelation and retrogradation. *Trends Food Sci. Technol.*, **1(1)**, 2–6.

Pomeranz, Y. (1985) *Functional Properties of Food Components*, Academic Press, New York.

Rapaille, A. (1986) Starch applications in the food industry, lecture presented at the *Symposium on Emulsifiers and Stabilisers for the Food Industry, Dublin, May 28*.

Rapaille, A. (1990) New technological trends for the production of starch based gums and jellies. *Confectionery Manufacture and Marketing*, June, pp. 51, 52.

Rapaille, A., Vanhemelrijck, J. and Mottar, J. (1988) The use of starches and gums in UHT milk desserts. *Dairy Industries International*, **53(9)**, pp. 21, 23, 25.

Whistler, R.L., Bemiller, J.N. and Paschall, E.F. (1984) *Starch: Chemistry and Technology*, 2nd edn, Academic Press, New York.

Cerestar Brochure on Food Starches. (1988) Cerestar SAINV, Avenue Louise 149, Bte 13, B-1050 Brusseis, pp. 4–9.

9 Xanthan gum

B. URLACHER and B. DALBE

9.1 Introduction

Xanthan gum was the first of a new generation of polysaccharides, produced by biotechnology. The polymer was discovered by the United States Drug Administration (USDA) and classified under the name B-1459 (xanthan gum). The gum, produced by *Xanthomonas campestris* NRRL B-1459, appeared to have valuable properties that would allow it to compete with natural gums. The production of xanthan gum started in the 1960s in the USA. Today there are four major suppliers worldwide and smaller manufacturers in Japan, Europe and the USA.

The advantages of a hydrocolloid obtained by fermentation are numerous: production and availability are not dependent on external factors such as weather, and a more consistent quality and performance of the texturising agent can be assured. Furthermore, the end price is not sensitive to economic or political problems.

In most countries food legislative authorities recognise xanthan gum as a valuable and harmless food additive. Where regulated, permitted use levels are in accordance with 'good manufacturing practice'. Xanthan gum has been approved in the USA since 1969 and in Europe since 1974.

Xanthan is registered in the European list of permitted thickening and gelling agents under the number E415 with a non-specified acceptable daily intake (ADI).

9.2 Process

Xanthan gum is a heteropolysaccharide produced by fermentation using the bacterium *Xanthomonas campestris*. This process is conducted using submerged aerobic fermentation in a sterile medium containing carbohydrates, a suitable nitrogen source, potassium phosphate and other trace minerals inoculated with the selected strain.

After a first inoculation with the selected strain in a pilot-scale fermenter, the incubation continues for 3 days at 30°C in an industrial-scale fermenter (Figure 9.1). The product is transferred to a holding tank and then a thermal treatment is applied to eliminate viable micro-organisms and to obtain a very low microbial count (Figure 9.2). The

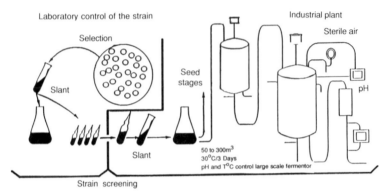

Figure 9.1 Industrial-scale production. Fermentation.

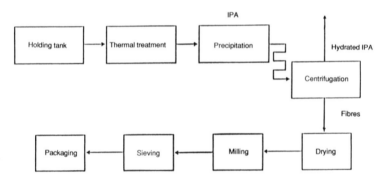

Figure 9.2 Industrial-scale production. Downstream processing.

product is then precipitated by isopropylalcohol (IPA) and the fibres are separated from IPA by centrifugation. These fibres are dried, milled and sieved before packaging.

Different products are manufactured:

- A standard grade with a particle size of 80 mesh (175 µm)
- A fine-mesh grade with a particle size of 200 mesh (75 µm), which is very rapid to hydrate but difficult to disperse and very dusty
- A granulated product with improved dispersability, especially in the absence of efficient stirrers, and which is non-dusty
- A transparent grade which provides clear solutions at low concentrations.

As xanthan gum is a sterilised product, the total microbial count of this type of gum is generally very low, much lower than that allowed by law.

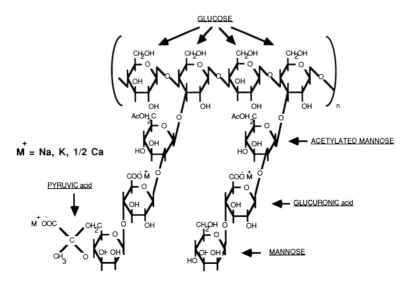

Figure 9.3 Primary structure of xanthan gum. (From Jansson *et al.*, 1975.)

9.3 Molecular structure

As shown in Figure 9.3 (Jansson *et al.*, 1975), the primary structure of the xanthan gum molecule is composed of a backbone of 1,4-linked β-D-glucose (like cellulose) with side chains containing two mannose and one glucuronic acid. According to Jansson *et al.* (1975), half of the terminal mannose units carry a pyruvic acid residue. These side chains represent a very large proportion of the molecule (about 60%) and give xanthan gum many of its unique properties. Xanthan gum has a high molecular weight of about 2 500 000 with a low polydispersity. Owing to the side chains, the polymer completely hydrates, even in cold water.

The secondary and tertiary structures are not so well characterised as the primary. The molecular conformation was interpreted by Moorhouse *et al.* (1977) using X-ray diffraction studies on xanthan fibres as being a helix (Figure 9.4) with a pitch of 4.7 nm (0.94 nm per disaccharide backbone). In this conformation it is possible that the molecule is stabilised through hydrogen bonds. Moorhouse *et al.* (1977) also proposed that the macromolecules in solution should be considered as a rigid helix. These authors did not reject the existence of a double or triple helix.

Xanthan in solution undergoes a conformational transition under the influence of temperature. This may indicate that xanthan gum goes from a rigid ordered state to a more flexible, disordered state (see Figure 9.5).

Milas and Rinaudo (1984; 1986) indicated recently that xanthan could

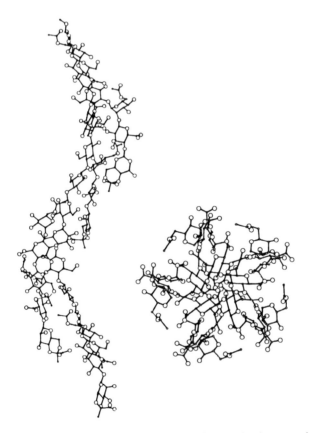

Figure 9.4 Molecular conformation of xanthan. (From Moorhouse *et al.*, 1977.)

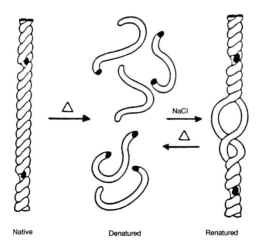

Native   Denatured   Renatured

Figure 9.5 Conformational transitions of xanthan gum in solution. (From Holztwarth and Prestridge, 1977.)

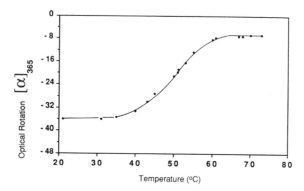

Figure 9.6 Thermal transition of xanthan gum measured by optical rotation (distilled water, concentration 0.25%).

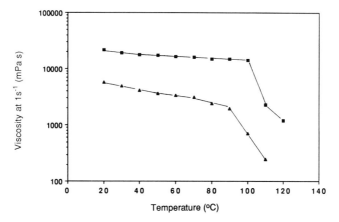

Figure 9.7 Thermal transition measured by viscosity in salt solution. ■, 1% xanthan gum in 1% sodium chloride solution; ▲, 0.5% xanthan gum in 1% sodium chloride solution.

have two ordered conformations: a native A conformation and a renatured B conformation. The B conformation has the same molecular weight as the native but a higher viscosity at the same concentration. The transition from native to denatured is irreversible whereas the transition from renatured to denatured is reversible. The transition temperature (T_m) depends on different factors, such as gum concentration and ionic strength, and also varies with pyruvic and acetic acid contents of the xanthan macromolecules.

This conformational transition can be measured by different analytical techniques such as optical rotation, calorimetry, circular dichroism and viscosimetry. The most useful and practical technique is optical rotation, as illustrated in Figure 9.6.

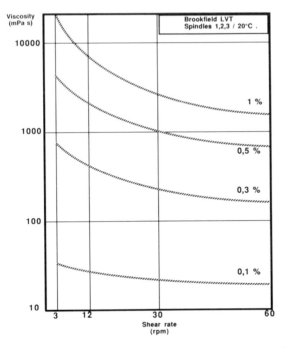

Figure 9.8 Effect of concentration and shear rate on xanthan gum solutions in distilled water.

The thermal transition temperature at low concentrations (0.1–0.3%) in distilled water is generally close to 40°C. In the presence of a small amount of salt and at concentrations generally used in food applications the thermal transition occurs at much higher temperatures, generally above 90°C (see Figure 9.7).

9.4 Xanthan gum in solution

9.4.1 Rheological properties

Aqueous solutions of xanthan exhibit a very high viscosity, even at low concentrations, and very strong pseudoplasticity with no evidence of thixotropy (Whitcomb and Macosko, 1978). These properties result from the unique rigid, rod-like conformation of xanthan in solution and from its high molecular weight: xanthan gum forms reversible entanglements at very low concentrations. In Figure 9.8 flow curves of xanthan solutions at different concentrations are presented: all solutions show a very high viscosity at low shear rates and a very strong pseudoplastic character,

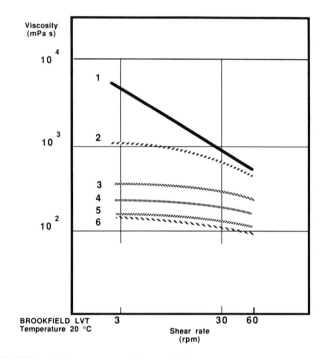

Figure 9.9 Effect of shear rate on different thickeners in distilled water (concentrations = 0.5% solutions made at room temperature). 1, xanthan gum; 2, guar gum; 3, hydroxyethylcellulose; 4, locust bean gum; 5, sodium carboxymethylcellulose; 6, sodium alginate.

which increases with concentration. This behaviour can have various advantages: as the viscosity decreases with the increasing shear rate, the product becomes easy to pour, mix or pump, and the organoleptic properties of food products are affected (the shear rate in the mouth is about $50\,s^{-1}$).

The thickening properties of xanthan compared with other food hydrocolloids are illustrated in Figure 9.9. Low shear rate viscosities show that values for xanthan solutions are always greater, especially at low concentrations. The shear-thinning character of xanthan solution is more pronounced than with other gums. This behaviour is due to the semi-rigid conformation of the xanthan polymer, which is more sensitive to shear than a random-coil conformation.

Another feature of xanthan gum solution is its viscoplasticity, which gives a high yield value, even at low concentrations. The yield value is the minimum shear stress required for a solution to flow. As illustrated in Figure 9.10, the yield value results from the formation of a weak network in the solution. This is the result of interactions between xanthan

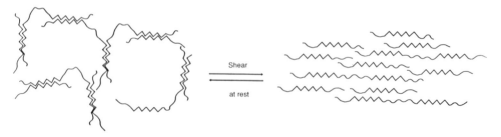

Figure 9.10 Weak network of xanthan macromolecules in solution.

Table 9.1 Yield values (mPa) of hydrocolloid at different concentrations in 1% KCl solution (Bingham extrapolation with a Rheomat 30 rheometer).

	Concentration (%)		
Hydrocolloid	0.3	0.5	1
Xanthan gum	500	2200	11 300
Guar gum	—	210	4000
Hydroxyethylcellulose	—	60	830
Locust bean gum	—	<50	360
Sodium carboxymethylcellulose	—	<50	410
Sodium alginate	—	<50	210

macromolecules, but the network is not a true gel because these interactions are not permanent and are totally shear reversible.

The yield value is difficult to measure because it is necessary to work at very low shear rates and frequently this value is extrapolated with different rheological models, such as those of Bingham and Herschel-Buckley (Launay *et al.*, 1986). In Table 9.1 it is clearly shown that xanthan is the only hydrocolloid to exhibit a significant yield value at low concentrations. This explains the ability of xanthan solutions to stabilise dispersions such as emulsions or suspensions. A good illustration of this stabilising property is given in Figure 9.11, in which settling rates of standardised particles of sand are compared for different hydrocolloids.

9.4.2 Stability and compatibility

Most food products contain salts, sometimes at a very high concentration (up to 15% in soya sauce), some foods are very acid, for example, dressings, fruit preparations and drinks, and many of them are heat treated using high-temperature short-time (HTST), ultra heat treatment (UHT) or sterilisation processes. Thus the stabilisers must be stable in various conditions of ionic strength, pH and temperature.

The secondary structure of the xanthan molecule in which the side

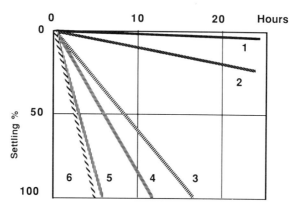

Figure 9.11 Suspending properties of silica sand (100–250 µm) in hydrocolloid solutions of 0.5% xanthan gum or 1% other gums in tap water. 1, xanthan gum (0.5%); 2, guar gum; 3, hydroxyethylcellulose; 4, locust bean gum; 5, sodium carboxymethylcellulose; 6, sodium alginate.

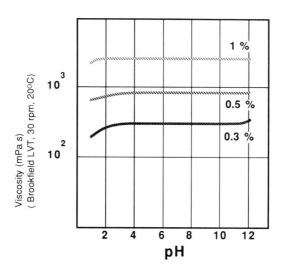

Figure 9.12 Effect of pH on solution viscosity at different xanthan gum concentrations.

chains are 'wrapped around' the cellulose backbone explains the unusual resistance of this hydrocolloid to degradation by acids or bases, high temperatures, freezing and thawing, enzymes and prolonged mixing.

9.4.2.1 Acids and bases. Xanthan solutions are stable over a wide range of pH. As shown in Figure 9.12, only extreme pH conditions (below 2.5 and above 11) affect properties of the solution. This stability is dependent on the gum concentration: the higher the concentration, the more stable

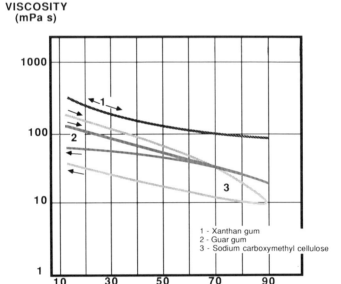

Figure 9.13 Temperature stability of xanthan gum compared with guar gum and sodium carboxymethylcellulose (concentration = 0.28% in tap water).

the solution. Xanthan can be used in formulations containing acetic, citric or phosphoric acid.

9.4.2.2 Temperature. Xanthan solution viscosity is only slightly affected by increasing the temperature from 10 to 90°C in the presence of salts which enhance heat stability (by stabilising the ordered conformation) (Figure 9.13). This property is quite unusual amongst the hydrocolloids. In food products, sterilisation treatments, such as 30 min at 120°C, are very common. In Figure 9.14 the stability of gum solutions is compared with xanthan gum; 90% of the initial viscosity is retained, whereas guar, alginates and carboxy methyl cellulose (CMC) show greater viscosity loss.

9.4.2.3 Enzymes. Xanthan is very resistant to enzyme degradation. It can be used in the presence of many common enzymes such as amylase, pectinase and cellulase, whether they originate from the raw materials or are specifically added during processing.

9.4.2.4 Freeze–thaw cycles and microwave treatment. Microwave treatment is direct, fast and selective heating. In most cases, even for stabilised products, microwave treatment causes moisture separation of the finished product, especially when a freeze–thaw cycle takes place. Consequently,

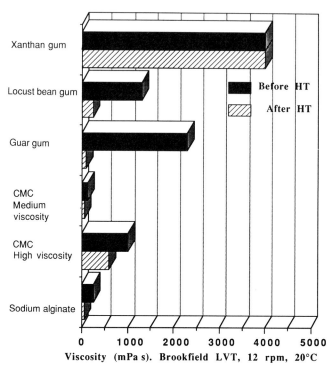

Figure 9.14 Comparison of heat stability of solutions containing different hydrocolloids at 0.6% in 2.5% sodium chloride heated at 120°C for 50 min. HT: heat treatment.

microwaveable foods need to be overstabilised, especially with a microwave-stable hydrocolloid.

Xanthan gum solution keeps all its viscosity after defrosting in a microwave oven, even at low concentrations (Figure 9.15).

9.4.2.5 Compatibility with salt and sugar. Although salts reduce the xanthan gum hydration rate, it is possible to obtain solutions with a high salt content that still have the desired rheological properties. At a gum concentration of around 0.4%, viscosity is unaffected by electrolytes, but around 1% there is a significant increase in viscosity in the presence of salts.

In the presence of sugar the hydration rate of xanthan is not modified, even at concentrations as high as 40–60% (Figure 9.16).

Xanthan is compatible with most of the ingredients in food formulations such as acids, salts, thickeners (starch, carrageenan, cellulose derivatives, gelatin and alginates) and proteins. However, interaction and precipitation may occur with some proteins, such as dairy proteins, if the system is acidic or heat processed.

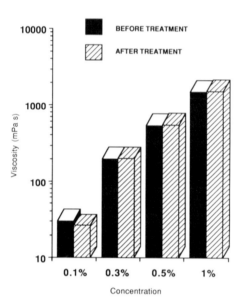

Figure 9.15 Microwave freeze–thaw stability of xanthan gum in 1% sodium chloride.

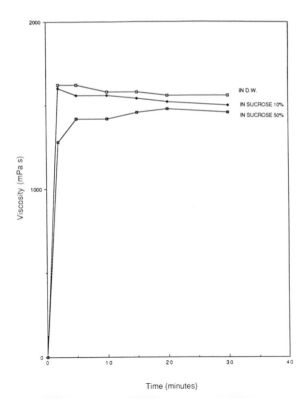

Figure 9.16 Hydration rate of xanthan gum in various sucrose concentrations (gum concentration 0.5%). DW: distilled water.

9.5 Solution preparation and use

Xanthan is a very hydrophilic product, and problems may arise during hydration if some basic rules are not respected. Before discussing the preparation of xanthan solutions, a distinction needs to be drawn between hydration and dispersibility:

- Dispersibility is the ease of separation of the individual gum particles when xanthan is introduced into the liquid
- Solubility is the ease with which these individual particles swell and develop viscosity.

It is necessary to find a compromise between dispersibility and hydration: an easily dispersible product will hydrate very slowly and vice versa. For example, a very fine powder is difficult to disperse, but once dispersed it is quick to hydrate. The hydration time depends on several factors:

- The effectiveness of the dispersion
- The size of the gum particles
- The other components of the formulation.

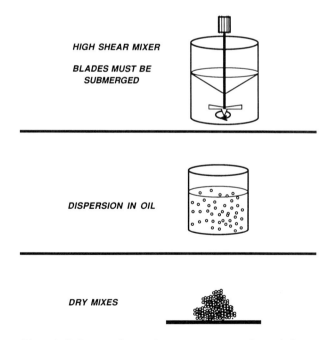

HIGH SHEAR MIXER

BLADES MUST BE SUBMERGED

DISPERSION IN OIL

DRY MIXES

Figure 9.17 Suggested procedures to prepare xanthan solutions.

Suggested procedures for the preparation of xanthan solutions without any lumping problems are shown in Figure 9.17.

1 Use a high-speed mixer (1500 r.p.m) if possible and slowly sprinkle xanthan onto the upper surface of the vortex.
2 If possible disperse xanthan with another component of the formulation such as:

   • a non-aqueous liquid, such as vegetable oil or ethyl alcohol, in which xanthan does not hydrate
   • other dry ingredients, such as sugar and flour.

For example, with salad dressings, good hydration can be obtained either by mixing xanthan gum powder with other dry ingredients (sugar, salt) or by dispersing the gum in vegetable oil.

On the industrial scale, rapid dispersion can be achieved by:

1 Dispersion funnels as shown in Figure 9.18: water rushes through the Venturi tube of the disperser and evenly draws xanthan from the funnel into the water by vacuum action.
2 The use of continuous process systems which produce colloidal solutions without entrapping air and which have a high process throughput. Powder and liquid phases are brought together in a cyclone chamber in metered amounts, and then perfectly dispersed and hydrated in an in-line flow section.

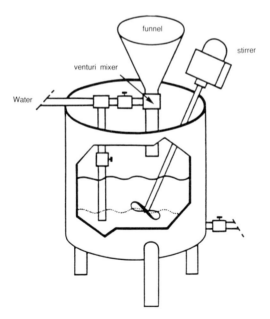

Figure 9.18 Dispersion funnel to prepare xanthan gum solutions.

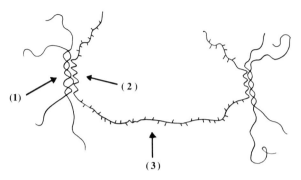

Figure 9.19 Interaction between xanthan gum and galactomannans. 1, xanthan macromolecules; 2, 'smooth' regions of galactomannan; 3, 'hairy' regions containing a high density of branched mannose units. (From Dea *et al.*, 1972.)

9.6 Gum associations

Many combinations of gums have proved interesting because their rheological and functional properties are complementary. In this area, gum combinations containing xanthan gum have not yet been fully exploited by the food industry. In addition, many specific and strong interactions occur between xanthan gum and different hydrocolloids. These interactions generally give a positive synergistic effect, such as enhancement of the viscosity or gelation, which is especially important and useful with galactomannans and glucomannans.

9.6.1 Xanthan–galactomannan interactions

All galactomannans, including guar gum, locust bean gum (LBG), tara gum and cassia gum, interact synergistically with xanthan gum. The best-known and commercially most valuable interactions are with guar gum and xanthan and between locust bean gum and xanthan.

Galactomannans are composed of mannose chains with galactose side units. The mannose to galactose ratio is an important parameter for the interaction with xanthan gum. Guar gum, which has a mannose to galactose ratio of around 2:1, exhibits weak synergism, whereas locust bean gum, with a ratio of about 4:1, reacts more strongly with xanthan gum.

The distribution of the galactose side chains in galactomannans is uneven and this synergistic effect is explained by the interaction between the xanthan gum molecule and the unsubstituted ('smooth') areas of the galactomannan (see Figure 9.19). Locust bean gum, which is much less branched than guar gum and which has a more favourable distribution

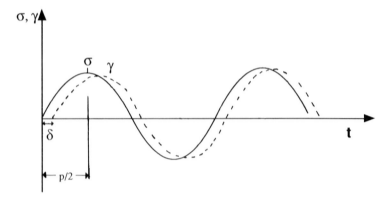

Figure 9.20 Phase diagram showing sinusoidal strain (σ) against stress (γ) with time. σ = phase difference.

of galactose units along the mannose chain, having considerably more 'smooth' regions, interacts more strongly. The exact conformation of xanthan gum involved in the mixed xanthan/LBG junction zones is still under debate (Cheetam and Mashimba, 1991). Some authors believe that the synergism can also be greatly affected by the pyruvate and acetate content of xanthan gum (Shatwell *et al.*, 1991; Tako, 1991).

9.6.1.1 Xanthan–guar interaction. The synergistic interaction between xanthan and guar gum is widely used in the food industry. This synergism is essentially an increase in viscosity and elastic modulus of the solution.

   In order to understand the importance of elastic modulus it must be realised that most polysaccharide solutions are not simply viscous liquids, but possess viscoelastic properties. These solutions have both properties of liquid (viscous) materials and properties of solid (elastic) materials. The viscoelastic properties can be determined by dynamic measurements. The principle is to apply a sinusoidal strain (σ) to the material and to measure the sinusoidal response (stress, γ). Both strain and stress are sinusoidal functions of time, with the same frequency but with a phase difference (δ) as shown in Figure 9.20.

   With a purely solid material the resulting stress is in phase with the strain, whereas with a purely liquid material there is deformation which cannot be recovered and is seen as the phase difference angle (called δ), which is equal to $p/2$. All other materials, such as xanthan gum solutions, have a phase difference angle which varies between 0 and $p/2$. These systems can be described by the measurement of the in phase modulus, G' (called elastic or storage modulus) and the out of phase modulus, G'' (called viscous or loss modulus). Another interesting rheological parameter is tan δ (with tan δ = G''/G'), which also gives information on the elasticity of the system. A low value for tan δ indicates an elastic

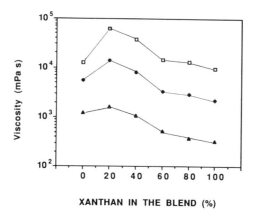

Figure 9.21 Viscosity of xanthan/guar blends versus blend ratio at different shear rates in distilled water at 1% total gum concentration. Shear rate: □, $0.2\,s^{-1}$; ♦, $1.6\,s^{-1}$; ▲, $21\,s^{-1}$.

network. (The relationships between these parameters are shown for xanthan gum–glucomannan gels in Figure 9.26).

The determination of the elastic modulus at a small strain level and a low rate of deformation provides information on the degree of elasticity of the solution and its ability to stabilise emulsions or suspensions. Knowledge of both parameters (apparent viscosity and elastic modulus) is important because some foods require a high elastic modulus to stabilise the emulsion or suspension (for example, salad dressings). In some other food applications a high apparent viscosity is required and a high degree of elasticity is undesirable, for example in some sauces or soups.

The variation in the apparent viscosity at different shear rates versus blend ratio is shown in Figure 9.21. As can be seen, the maximum viscosity synergism is achieved for blends containing a low level of xanthan gum (about 20%, w/w). This study was performed in distilled water and at a total gum concentration of 1%. Unfortunately, the synergism strongly decreases in the presence of salts and, consequently, the viscosity synergism is generally lower in most food applications. The synergism is also dependent on the total gum concentration: the higher the gum concentration, the higher the synergism.

The variation in elastic modulus at a low rate of deformation (0.158 rad/s) with blend ratio is shown in Figure 9.22. In a salty medium and at 1% total gum concentration, all blend ratios have lower values than xanthan gum alone, but blends containing up to 30% guar gum possess rheological properties similar to those of pure xanthan gum. Thus, some xanthan/guar blends have very valuable applications as stabilisers.

Another advantage of using xanthan/guar gum blends is an improvement in the thermal stability of guar gum. As shown in Figure 9.23, the

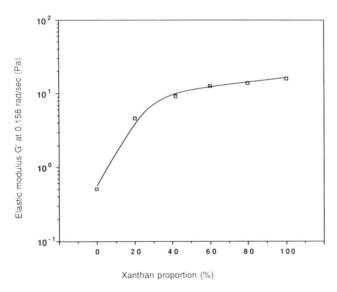

Figure 9.22 Elastic modulus (G') of xanthan/guar blends at 1% total gum concentration in 1% potassium chloride solution.

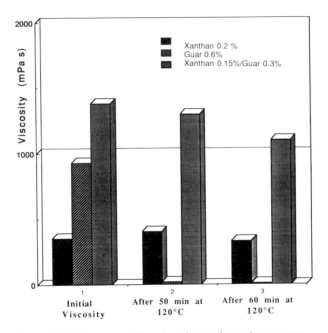

Figure 9.23 Thermal stability of xanthan and guar in tap water.

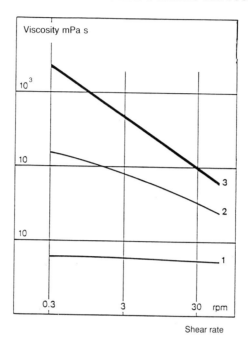

Figure 9.24 Viscosity of LBG/xanthan solutions at 0.1% total gum concentration using Brookfield LVT at 20°C. 1, locust bean gum; 2, xanthan gum; 3, locust bean gum/xanthan gum (50:50).

thermal stability of guar is dramatically increased in the presence of xanthan, but the thermal stability of all blends is lower than that of xanthan gum alone.

In summary, because a small proportion of xanthan in guar gum dramatically alters the rheological properties, the xanthan/guar blends are widely used in the food industry as very efficient thickening agents. Compared with xanthan alone, these blends are generally less efficient in terms of stabilising emulsions and similar products and providing good thermal stability, but because of their smooth, even flow properties and their lower cost they have very widespread application in foods.

9.6.1.2 Xanthan–LBG interaction. As mentioned above, the interaction between xanthan and LBG is strong and it is possible, above a certain concentration, to obtain gels.

At low concentrations there is a strong increase in the yield value and viscosity (Figure 9.24). At concentrations above about 0.2% xanthan–LBG combinations produce strong, cohesive, thermoreversible gels with very low syneresis (Figure 9.25). The optimum ratio is between 40:60 and 60:40. However, because the xanthan/LBG mixed gel is so elastic,

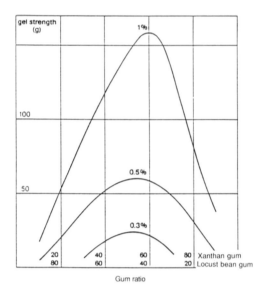

Figure 9.25 Effect of the gum ratio on the strength of xanthan/LBG gels, measured with a Stevens–LFRA texture analyser at 20°C.

cohesive and non-brittle, it is not pleasant to eat. The texture of the mixed gel can be improved by the addition of another biopolymer, such as starch or protein, but there are few food applications for this kind of gel. For the moment this gelling system is only widely used in pet food products.

Because of the yield value of xanthan/LBG solutions, stable suspensions can be attained at a very low gum concentration. The stability of the blend under different conditions of pH, temperature and salt is lower than xanthan but higher than LBG alone. Unfortunately this very effective combination is especially difficult to control because of the instability between the solution and gel states.

9.6.2 Xanthan–glucomannan (konjac mannan) interactions

Xanthan interacts with glucomannan (konjac mannan) in a very similar way to its interaction with LBG. The synergism is very strong and provides gels at low concentration above about 0.2%. The optimum ratio is between 40:60 and 30:70 xanthan–konjac mannan. However, the interaction is decreased in the presence of salt. Typical thermal behaviour of these thermoreversible gels is illustrated in Figure 9.26. At temperatures below 50°C the system shows well-defined elastic properties, the storage modulus G' being greater than the loss modulus G''. At about 55°C, a critical temperature is reached with collapse of the gel structure and a

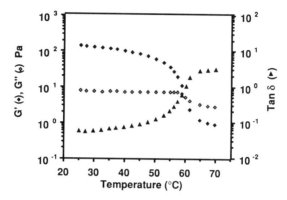

Figure 9.26 Thermal behaviour of a xanthan/glucomannan gel. Xanthan: konjac = 30:70; total gum concentration 0.5%, measured with Carrimed rheometer; temperature sweep 2.25°C/min; torque 150 µN m; frequency 1 Hz. (From Dalbe, 1992.)

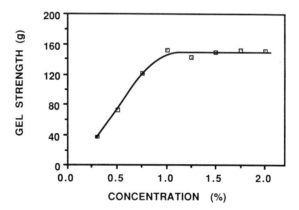

Figure 9.27 Variation in gel strength with total gum concentration for konjac mannan/xanthan gum blend (70:30) using a Stevens–LFRA texture analyser at 22°C. (From Dalbe, 1992).

sharp decrease in G'. Above this temperature, the behaviour is typical of a liquid with G'' significantly higher than G'.

The variation in gel strength with the total gum concentration is shown in Figure 9.27. Maximum gel strength is reached at about 1% total gum concentration and further increases in concentration do not give additional increase in gel strength. An identical phenomenon of a gel strength plateau has been observed with LBG–xanthan gels.

The mechanical properties of xanthan–konjac gels are somewhat similar to those of xanthan–LBG and, likewise, these gels are also unpleasant to eat.

9.6.3 Xanthan–starch interactions

As starch is the most commonly used hydrocolloid, there is great interest in improving its storage properties, including preventing retrogradation, water release and stability.

Xanthan gum has little effect on the gelatinisation of starch but considerably improves its properties. Small amounts of xanthan help to stabilise starch solutions during storage, even at low gum concentrations. Solutions of starch at 1.5 or 3% (native or modified) are not stable at room temperature, but adding 0.1% or 0.2% xanthan prevents starch from retrograding and makes the solution stable to storage. Starch is not stable at low pH but 0.1–0.2% xanthan can stabilise the starch paste at pH around 3. Starch is not stable after a freeze–thaw cycle, especially if the product is thawed in a microwave oven. However, 0.1 or 0.2% xanthan can make these solutions stable to such treatments.

In summary, in many cases, small amounts of xanthan improve the stability of starch when it is the basic thickener of the formulation.

9.7 Applications

9.7.1 Dressings

In dressings the first objective is to stabilise the oil–water emulsion for periods of storage as long as 1 year. The ideal product for this type of application should provide a high yield value for good emulsion stabilisation and strong pseudoplasticity to facilitate the manufacturing operations of mixing, pumping and filling, and give finished products which flow easily. Xanthan gum, which provides a high viscosity at rest and a high pseudoplasticity, has a lot of advantages for this application. The stability of the emulsion is not affected by the pH (around 3.5 in some dressings), salts (as high as 15% in barbecue sauces) or thermal treatments (UHT or pasteurisation). Another advantage is that xanthan gum exhibits a quite uniform viscosity between 5 and 70°C which provides the product with good stability and texture in different storage conditions. The high yield value of xanthan makes it possible to suspend spices, herbs and vegetables in the product, allows the salad dressing to cling to the salad and to appear to have 'body'. The level of use is dependent on the oil content in the product:

- About 0.2–0.3% in a high-oil formulation (50–60% oil)
- About 0.3–0.4% in a medium-oil formulation (around 30% oil)
- About 0.4–0.6% in a low-oil product (10–20% oil).

As shown in Figure 9.28, it is possible to obtain the same flow properties with different levels in oil by adjusting the level of xanthan.

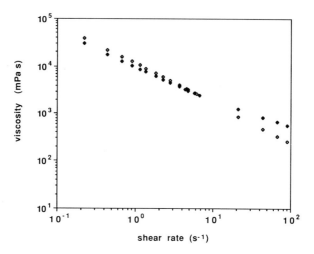

Figure 9.28 Flow curves of dressings with different oil content. $\Diamond$, 10% oil, 0.8% xanthan; $\blacklozenge$, 55% oil, 0.25% xanthan.

9.7.2 Sauces, gravies, relishes and canned soups

Even at low concentrations, xanthan gum imparts a high viscosity to sauces and gravies. In these products the viscosity is maintained across a wide temperature range and is resistant to different formulation changes. Sauces and gravies stabilised with xanthan gum are especially resistant to thawing and heating in a microwave oven. The pseudoplasticity of xanthan gives sauces and gravies a clean mouthfeel and good flavour release.

9.7.3 Dairy products

In dairy products, a small amount of xanthan gum (0.02–0.05%) may be used in combination with other hydrocolloids including carrageenan, guar gum and locust bean gum. In ice cream the role of xanthan is essentially to control ice crystal formation, give slow meltdown and modify and improve some textures. A small amount of xanthan with carrageenan in milk gels reduces the brittleness and syneresis of this gel. In the latter application, the level of use is between 0.05 and 0.2% depending on the gelling system used and the texture required.

9.7.4 Whipped creams and mousses

The high yield value of xanthan provides good stabilisation of the air cells in whipped products. Furthermore, the whipping process is easier because of the high pseudoplasticity of xanthan gum. In addition, with xanthan or xanthan/locust bean gum combinations, the stability of whipped creams can be maintained even in contact with other food ingredients, such as bakery fillings.

9.7.5 Instant mixes: drinks, soups and desserts

Because of its very fast hydration in different media, especially at room temperature, xanthan gum is very effective in instant mixes. Its role is essentially to thicken, suspend and give body to the product. The concentration of xanthan in the final food product is between 0.1 and 0.2%. In instant beverages xanthan gum can be used in combination with sodium carboxymethylcellulose or guar gum.

9.7.6 Bakery products

Owing to the pseudoplasticity of xanthan gum, processing operations on dough, which include pumping, kneading and moulding, are easier to achieve when low levels of gum, between 0.05 and 0.2% are included in formulations.

Because of its water retention properties, xanthan gum prevents lump formation during kneading and improves the homogeneity of the dough. Furthermore, xanthan reduces the loss of water during cooking and storage of the final food product.

Another advantage of xanthan gum is that the volume of bakery products, such as sponge cake, is greater and the distribution and the size of the air cells are more uniform.

9.7.7 Syrups, toppings and fillings

Xanthan gum can be used in chocolate syrups in order to maintain cocoa particles in suspension. In this particular application, a small amount of xanthan (0.05–0.1%) is sufficient to obtain good stabilisation and does not affect the texture of the product itself. In toppings and fillings, xanthan improves texture and helps to improve the freeze–thaw stability and control syneresis.

9.7.8 Pet foods

In gravies, xanthan in combination with guar is an excellent thickener and stabiliser and gives smooth-textured products. The good thermal stability of the blend ensures the sauce keeps its high viscosity even after very strong sterilisation treatment.

In gelled products, combinations of xanthan and locust bean gum or locust bean gum and carrageenan produce very efficient gelling systems. As xanthan maintains its viscosity at high temperatures, its stabilising ability is not affected during the sterilisation step and it provides very homogeneous products after cooling.

References

Cheetam, N.W.H. and Mashimba, E.N.M. (1991) *Carbohydr. Polym.*, **14**, 17–27.
Dalbe, B. (1992) Interactions between xanthan gum and konjac mannan. In: *Gums and Stabilisers for the Food Industry*, Vol. 6, G.O. Phillips, D.J. Wedlock and P.A. Williams, eds, Oxford University Press, Oxford, pp. 201–208.
Dea, I.C.M., McKinnon, A.A. and Rees, D.A. (1972) *J. Mol. Biol.*, **68**, 153–172.
Holztwarth, G. and Prestridge, E.B. (1977) *Science*, **197**, 757–759.
Jansson, P.E., Kenne, L. and Lindberg, B (1975) *Carbohydr. Res.*, **45**, 275–282.
Launay, B., Doublier, J.L. and Cuvelier, G. (1986) In: *Functional Properties of Food Macromolecules*, J.R. Mitchell and D.A. Ledward, eds, Elsevier Applied Science Publishers, UK, pp. 1–78.
Shatwell, K.P., Sutherland, I.W., Ross-Murphy, S.B. and Dea, I.C.M. (1991) *Carbohydr. Polym.*, **14**, 29–51.
Tako, M. (1991) *Carbohydr. Polym.*, **16**, 239–252.
Whitcomb, P.J. and Macosko, C.W. (1978) *J. Rheol.* **22(5)**, 493–505.

10 Gellan gum

W. GIBSON

10.1 Introduction

Gellan gum is the generic name for the extracellular polysaccharide elaborated by the bacterium *Pseudomonas elodea*. Because of its gel-forming properties, it has potential use in the food industry. The polysaccharide, which can be produced in either a substituted or an unsubstituted form, produces gels at low concentration when hot solutions of the gum are cooled. The substituted form produces soft, elastic gels, whereas the unsubstituted form produces hard, brittle gels.

10.2 Manufacture

Gellan gum, proprietary to Kelco, (Kang *et al.*, 1980; 1982; 1983; Kang and Veeder, 1982; 1983) is manufactured in an aerobic, submerged, fermentation process (Figure 10.1) (Kang *et al.*, 1981). A pure culture of *Pseudomonas elodea* is inoculated into a fermentation medium consisting of a carbon source, such as glucose, a nitrogen source and a number of inorganic salts. Product consistency is assured by strict control of the fermentation conditions such as pH, temperature, aeration and agitation. The fermentation broth becomes increasingly viscous as the organism metabolises the glucose and gellan gum is secreted. When the carbon source is exhausted, this viscous broth is pasteurised to kill off viable cells before the gum is recovered. Treatment of the pasteurised broth with alkali removes the acyl substituents on the gellan gum backbone. Following removal of the cellular debris, the gum is recovered by precipitation with alcohol. This produces the unsubstituted form of gellan gum with a high degree of purity.

10.3 Chemical composition

Gellan gum is a linear, anionic heteropolysaccharide with a molecular weight of around 0.5×10^6 daltons. It is composed of tetrasaccharide repeat units (Figure 10.2) comprising 1,3-β-D-glucose, 1,4-β-D-glucuronic acid, 1,4-β-D-glucose and 1,4-α-L-rhamnose. The polymer, as secreted by

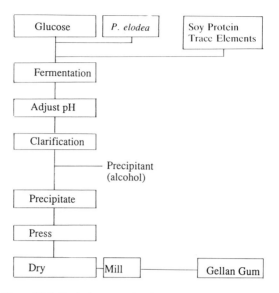

Figure 10.1 Typical process for the manufacture of gellan gum.

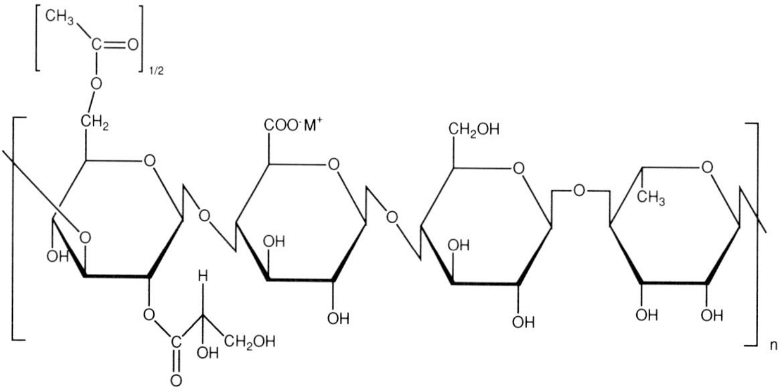

Figure 10.2 Substituted tetrasaccharide repeating unit of gellan gum.

the micro-organism, contains approximately 1.5 acyl substituents per tetrasaccharide repeating unit. These have been identified as an L-glyceric ester on C-2 of the 3-linked D-glucose and an acetic ester on C-6 of the same glucose residue (Kuo *et al.*, 1986). The presence of these substituents, in particular the bulky glycerate groups, hinders chain association and accounts for the change in gel texture brought about by de-esterification.

The de-esterified product is a polymer with a well-defined, unsub-

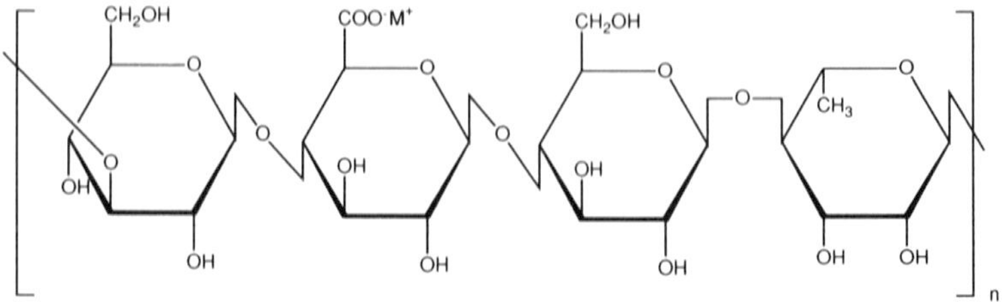

Figure 10.3 Unsubstituted tetrasaccharide repeating unit of gellan gum.

stituted, tetrasaccharide repeating unit (Figure 10.3). In the solid state the molecule forms a parallel, half-staggered intertwined double helix in which each polysaccharide chain is a left-handed, threefold helix (Chandrasekaran *et al.*, 1988a,b).

The presence of uronic acid residues in the structure means the polysaccharide can be presented as a variety of salts. As produced, it exists as a mixed salt, predominantly in the potassium form but also containing small amounts of sodium, calcium and, to a lesser extent, magnesium.

10.4 Functional properties

10.4.1 Hydration

The presence of divalent ions in the polysaccharide has an inhibiting effect on its hydration. In cold deionised water, only partial hydration is obtained. It is necessary to heat the dispersion to at least 70°C to achieve complete hydration. The inhibiting effect of the divalent ions is confirmed by the fact that complete hydration can be achieved, in cold deionised water, using a pure monovalent salt form of the gum.

The presence of divalent cations in most water supplies further restricts hydration at ambient temperature. Therefore, in most practical situations gellan gum can be dispersed easily without hydrating and, consequently, without the problems of lumping normally associated with cold water-soluble gums. The extent to which hydration takes place in cold water depends upon the cation concentration.

In soft water, sufficient hydration of the gum can take place to present problems of agglomeration and lumping. In such circumstances it is best to dry blend the gum with sugar as a dispersion aid, or to use good mechanical agitation to improve dispersion.

In hard water (above 150 ppm calcium carbonate) some hydration takes place at ambient temperature, but it is insufficient to create dispersion

Table 10.1 Temperature at which 0.25% gellan gum dissolves in water of varying degrees of hardness.

Water hardness (ppm calcium carbonate)	Hydration temperature (°C)
0	75
100	90–95
200	>100

problems. In these circumstances, gellan gum is easily dispersed by hand stirring. To achieve full hydration of the gum the dispersion must be heated. The temperature at which the gum hydrates depends on the cation concentration: the greater the cation concentration, the higher the temperature required for full hydration (Table 10.1).

In water hardness of above about 200 ppm calcium carbonate, gellan gum will not fully hydrate even in boiling water, but in processes which permit even higher temperatures to be achieved, for example in a retort or ultra heat treatment (UHT) process, complete hydration can be achieved in even higher concentrations of cations.

The presence of monovalent ions can also inhibit the hydration of gellan gum, but the levels necessary are considerably higher than those of the divalent ions. Thus, the latter may be sequestered with compounds such as sodium or potassium citrate or various sodium or potassium phosphates, enabling gellan gum to be dissolved at lower temperatures, without introducing sufficient monovalent ions to interfere with hydration. It is even possible to hydrate gellan gum fully at ambient temperature with the use of sufficient sequestrant. By carefully controlling the sequestrant concentration it is possible to choose the temperature at which gellan gum will fully dissolve (Figure 10.4).

10.4.2 Solution properties

When dissolved in cold water with the help of a sequestrant, gellan gum at 1% concentration produces a highly viscous solution, which is less pseudoplastic or shear thinning than a xanthan gum solution, but more pseudoplastic than an high molecular weight sodium alginate solution (Figure 10.5). The gellan gum solution is sensitive to increases in temperature and undergoes a dramatic drop in viscosity over a very small temperature range at a relatively low temperature (Figure 10.6). This change in viscosity, which has been monitored over the range 25–50°C, is completely reversible and most probably reflects a conformational change from some form of relatively ordered, non-aggregated double helix to a random coil, as suggested by Robinson *et al.* (1987). The practical sig-

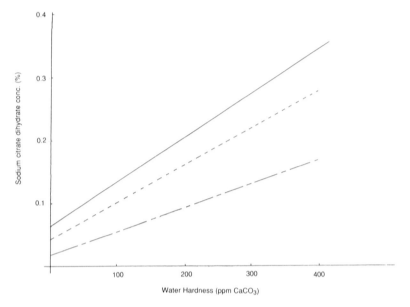

Figure 10.4 Concentration of sodium citrate required to hydrate gellan gum. —, 30°C; ---, 50°C; ——, 70°C.

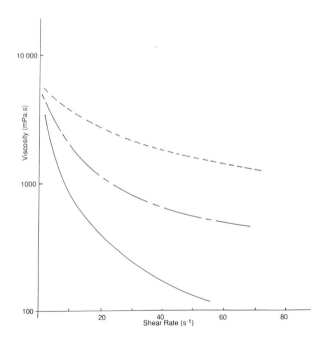

Figure 10.5 Comparison of viscosity versus shear rate. —, xanthan gum; ——, gellan gum; ---, sodium alginate.

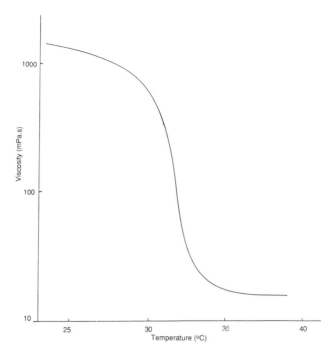

Figure 10.6 Effect of temperature on viscosity of a 1% gellan gum solution, shear stress 20 N/m².

nificance of this large drop in viscosity on heating is that a stock solution can be prepared at a relatively high concentration without it becoming too viscous to handle.

Like other polysaccharides, gellan gum will undergo hydrolytic degradation at elevated temperatures, especially in acid conditions. Therefore prolonged heating of solutions at high temperatures could result in reduction of the gel strength of the final gels. Under acid conditions, degradation is accelerated, but even at pH 3.5 a gellan gum solution can be maintained at 80°C for up to 1 h with only minimal deterioration in the quality of the final gel. However, under neutral conditions, gellan gum is remarkably stable, and may be held for several hours at up to 80°C without a significant change in final gel quality.

10.4.3 Gel formation

The mechanism whereby gellan gum forms gels in water is not fully understood, but many authors have suggested that gelation initially occurs by the formation of double helices followed by ion-induced association of these double helices. A recent model proposed by Gunning and Morris

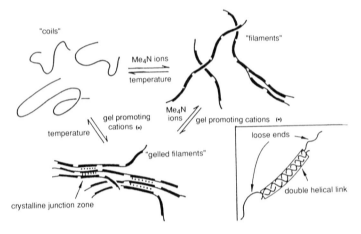

Figure 10.7 Model for the gelation of gellan gum. (After Gunning and Morris, 1990.)

(1990) suggests that heating and cooling in the absence of gel-promoting cations favours the formation of fibrils by double helix formation between the ends of neighbouring molecules (Figure 10.7). In the presence of gel-promoting cations these fibrils associate, with the formation of gels.

Gelation is dependent upon the ionic strength and identity of the cation. Divalent cations are more effective than monovalent cations and typically produce gels at around 3–7% of the required level of monovalent cations.

Solutions of gellan gum prepared at ambient temperature will react with additional mono- or divalent cations to form gels. Homogeneous gels are formed by the controlled release of divalent cations into the polysaccharide solution using the same techniques used in alginate gel formation. However, gels prepared in this way with gellan gum tend to be unstable and exhibit syneresis.

Of much more practical value are the gels which are prepared by cooling a hot solution of the gellan gum. A dispersion of gellan gum, with the minimum amount of sequestrant needed to hydrate the gum fully at an elevated temperature such as 80–90°C, will form a coherent, demouldable gel on cooling to ambient temperature (Table 10.2). For optimum gel strength however, it may be necessary to add more divalent cations. This may be achieved by adding a cationic salt solution to the hot gum solution before cooling. In this way, demouldable gels can be prepared with gellan gum at concentrations as low as 0.05%.

The gelation temperature for gellan gum solutions depends on the cation concentration and, typically, increases from 35 to 55°C with increasing cation concentration. Provided solutions are maintained above the setting temperature, gel formation will be delayed. Like other gel

Table 10.2 Gel formation with gellan gum. (After Sanderson *et al.*, 1989.)

Water hardness (ppm calcium carbonate)	Added sodium citrate (%)	Hydration temperature (°C)	Added Ca^{2+} (%)	Gel strength (N/cm^2)
0	0	75	0.024	5.1
100	0.05	25	0.032	5.4
300	0.10	65	0.024	5.1
600	0.20	65	0.024	5.0

systems it is preferable, after the onset of gelation, to leave the system undisturbed while the gel matures.

10.4.4 Gel properties

Depending on the degree of esterification, gellan gum will form gels with a range of textural properties from soft and elastic at one extreme to hard and brittle at the other. Since, at the time of writing, only the de-esterified product is commercially available, only the hard and brittle gellan gum gels will be considered.

One of the most important features of a gelling agent in food is the texture it provides. A technique known as texture profile analysis (TPA) has been used to describe quantitatively the texture of gellan gum gels and to measure the effects of a number of variables. The technique uses an Instron universal testing machine interfaced with a computer to compress a gel specimen twice in succession. From the force–deformation curve produced, the computer calculates values for the following parameters:

Hardness. The maximum force which occurs during the first compression cycle, usually corresponding to the rupture strength of the gel.

Modulus. The perceived firmness when the gel is squeezed by a small amount. It is analogous to the gentle squeezing of fruit to determine its ripeness.

Brittleness. A measure of how far the gel can be compressed before it ruptures. A gel that ruptures early in the compression cycle is more brittle than one that cracks later, therefore a low number is indicative of high brittleness.

Elasticity. A measure of how much the initial gel structure is broken down during the first compression cycle. A gel that is broken only by a small amount will recover more than one that is more broken, therefore a high number is indicative of a high degree of elasticity.

A more detailed account of TPA is given by Sanderson *et al.* (1987a).

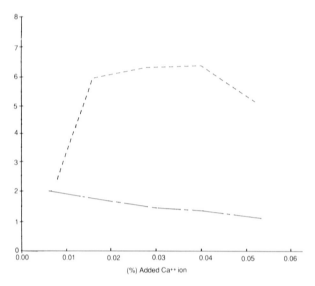

Figure 10.8 Effect of calcium ion concentration on the hardness and modulus of 0.25% gellan gum gel. (After Sanderson *et al.*, 1987.) – – –, modulus (N/cm²); — –, hardness (kgf).

10.4.4.1 Effect of cation concentration. Figure 10.8 shows the hardness and modulus of gellan gum gels (0.25%) as a function of calcium ion concentration under neutral conditions. Gels were prepared by dispersing gellan gum in distilled water, heating to 80°C to hydrate the gum before adding the appropriate amount of calcium ions as a calcium chloride solution, and cooling. In the absence of added cations, only a very weak gel is formed, but the hardness rapidly increases to a maximum at very low calcium concentrations and gradually decreases as the ionic concentration increases. The modulus shows a somewhat symmetrical increase and decrease with increasing calcium content. These strength parameters change very little over the calcium ion range 0.016–0.05%. Therefore, provided products are formulated so that the free calcium ion concentration is in this range, the gel strength will be fairly insensitive to changes in the ionic composition caused by variations in the ionic content of other food ingredients. It is also worth noting that the lower end of this calcium ion range (0.016%) equates to 400 ppm as calcium carbonate. Since this is higher than the hardness of most water supplies, some products will require additional calcium ions to achieve optimum results.

Figure 10.9 shows the characteristic brittle texture of gellan gum gels with values in the range 30–40%, over the calcium ion concentration range 0.01–0.05%. Gels tend to be a little more brittle at the higher ionic concentration. In contrast, elasticity is highest at low calcium ion concentration and reduces rapidly to an almost constant value of around

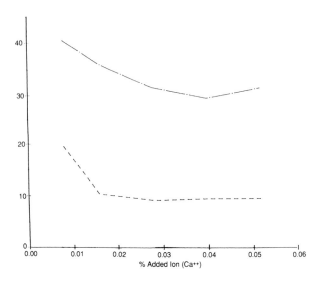

Figure 10.9 Effect of calcium ion concentration on brittleness ($\cdot$—$\cdot$, %) and elasticity (– – –, %) of 0.25% gellan gum gel. (After Sanderson *et al.*, 1987.)

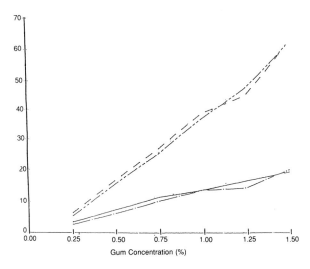

Figure 10.10 Effect of gum concentration on gellan gum gel hardness and modulus at a calcium ion concentration of 0.03%. (After Sanderson *et al.*, 1987.) —, hardness, pH 7.5 (kgf); – – –, modulus, pH 7.5 (N/cm^2); —$\cdot$—$\cdot$, hardness, pH 4.0 (kgf); – – – – –, modulus, pH 4.0 (N/cm^2).

10% with increasing cation concentration. Like the strength parameters, these flexural parameters change very little over the calcium ion range 0.016–0.05%.

Other divalent ions have similar effects on gel texture; magnesium, in particular, has the same effect as calcium at equal ionic concentration. These trends are also obtained with monovalent ions, but at much higher concentrations. For example, the maximum hardness with sodium or potassium ions is achieved at an ionic concentration about 25 times greater than the molar concentration of calcium or magnesium ions. At the low concentrations of gellan gum likely to be used in food applications, sodium ions have practically the same effect as potassium ions on gel strengh. At high gum concentrations, Grasdalen and Smidsrod (1987) have shown that potassium ions produce stronger gels than sodium ions at the same concentration.

10.4.4.2 Effect of gum concentration. It is not surprising that the strength of gellan gum gels increases with increasing concentration. However, compared with agar or κ-carrageenan gels (also prepared under optimum ion concentration) gellan gum gels are stronger at equal gum concentration (Figure 10.10). While the differences in gel hardness alone show gellan gum to be a more efficient gel former, the differences in gel modulus reveal even greater efficiency. For example, at 0.5% gum concentration, gellan gels are nearly eight times firmer (higher modulus) than either agar or κ-carrageenan gels. It is the high modulus value that gives gellan gum gels the ability to stand freely with little or no sag.

Unlike hardness and modulus, which are strongly dependent on gum concentration, the flexural parameters do not change significantly (Figure 10.11). Brittleness shows a small but insignificant increase of about 6% over a fourfold increase in gum concentration. Elasticity nearly

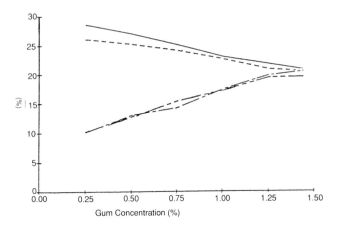

Figure 10.11 Effect of gum concentration on gellan gum gel brittleness and elasticity at a calcium concentration of 0.03%. (After Sanderson *et al.*, 1987.) —, brittleness, pH 7.5; ----, brittleness, pH 4.0; — —, elasticity, pH 7.5; — — — elasticity, pH 4.0.

Table 10.3 Effect of sucrose on gel texture.

Sucrose concentration (%)	Hardness (kg f)	Modulus (N/cm^2)	Brittleness (%)	Elasticity (%)
0	4.3	12	31.4	16.8
20	4.7	12.5	31.3	16.2
40	6.4	15	34.7	19.9
60	6.0	2.2	58.1	40.0

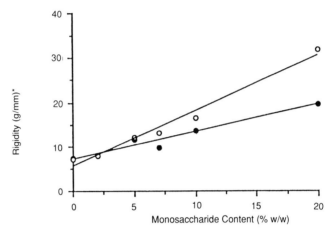

Figure 10.12 Effect of sugar composition at 60% solids on strength of 0.5% gellan gum gel. ○, fructose; ●, glucose; *, ratio of load required for gel to rupture to distance penetrated to rupture.

doubles over the same gum concentration range, to a value of 18%, but a much larger increase would be required to produce a change in gel character that would be detectable on eating.

10.4.4.3 Effect of pH. The texture of gellan gum gels is remarkably constant over the pH range encountered in many food systems. Reducing the pH from 7.5 to 4.0 by the addition of citric acid causes insignificant changes in product eating quality over a wide range of gum concentration.

10.4.4.4 Effect of sugars. Sucrose, a major ingredient in dessert and confectionery products, can have a pronounced effect on gel texture (Table 10.3). Up to 20% sucrose has no significant effect on gellan gum gel texture. Even at up to 40% concentration, sucrose has very little effect. However, at concentrations in excess of 40%, sucrose has a marked effect on the texture of gellan gum gels. For example, at a

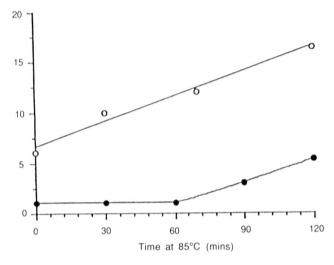

Figure 10.13 Effect of hot storage of a gellan gum solution in 60% sucrose at pH 3.4 on ultimate gel strength. ○, rigidity (g/mm; ratio of load required for gel to rupture to distance penetrated to rupture); ●, reducing sugar (%).

concentration of 60% sucrose, the gel remains hard, but becomes considerably less firm (lower modulus), less brittle and more elastic. These findings will be of importance in confectionery applications in which the solids content is even higher, albeit with a mixture of sugars.

Solubility limitations preclude testing glucose at 60% concentration. However, it has been shown that partial replacement of the disaccharide, sucrose, by either of the monosaccharides, glucose or fructose, produces firmer, harder gels than sucrose alone at 60% total concentration (Figure 10.12). Results suggest that at high concentrations the smaller sugar molecules may interfere less than sucrose with the gelation of gellan gum. Indeed, there may be some enhancement of the gel strength by smaller sugar molecules. This is borne out in an experiment carried out to monitor the degradation of gellan gum in 60% sucrose solution at pH 3.4 at 85°C (Figure 10.13). Over a 2 h period, contrary to an expected decrease in resultant gel strength on cooling, it was found that the gel strength increased. This is attributed to the inversion of some of the sucrose to glucose and fructose during the heating period and the fact that the increase in gel strength resulting from a change in sugar composition was greater than the reduction in gel strength arising from gellan gum degradation.

10.4.4.5 Setting and melting points. Key properties for any gel are its setting and melting points. For gellan gum gels these temperatures

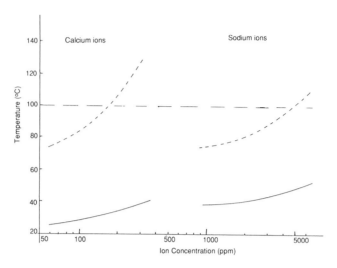

Figure 10.14 Effect of cation concentration on setting and melting temperatures of 0.2% gellan gum gels. —, setting point; – – –, melting point.

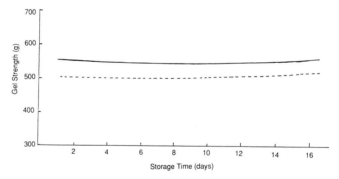

Figure 10.15 Storage stability of 0.25% gellan gum gels at pH 7.5. – – –, stored at 20°C; —, stored at 4°C.

depend primarily on the identity and concentration of cations and, to a lesser extent, on the gum concentration.

Gels made with calcium ions set at about 25–40°C at calcium concentrations in the range 40–400 ppm, while those made with sodium ions at concentrations of 1000–5000 ppm set at the slightly higher temperatures of 35–50°C (Figure 10.14). Ionic concentration has a marked influence on melting temperature, which is at least 40°C higher than the setting temperature. At lower ion levels, gels tend to remelt at temperatures between 75 and 100°C, while at the higher ion concentrations gels do not melt below 100°C. This property is useful in applications where thermostable gels are required.

10.4.4.6 Gel stability. The setting time of a gellan gum gel is largely dependent on the rate of cooling. Once the setting temperature has been reached, the gel quickly reaches a mature strength which is slightly lower at ambient temperature than under refrigeration (Figure 10.15). When stored under these conditions gels show no signs of toughening, with gel strength remaining essentially constant. This stable gel strength confirms observations that gellan gum gels do not undergo syneresis, as syneresis is normally accompanied by increased gel strength as a result of gel shrinkage.

However, at very low gum levels, free water can be expressed from the gel by pressure. At a gellan gum concentration of 0.1–0.2% water can be squeezed from the gel. If the unit volume of the gel is large enough, a demoulded, free-standing gel may undergo syneresis under its own weight. Therefore, for products intended to be demoulded, it may be necessary to use higher gellan gum concentrations (typically >0.2%) or to include thickeners to suppress potential syneresis.

10.4.4.7 Sensory properties. Gellan gum produces clear gels which are firm to touch, yet their characteristic brittleness makes them easy to break down. On eating, the gel disintegrates relatively quickly so that, although the gel does not melt at body temperature, water released during mastication gives a melt-in-the-mouth sensation. This release of water on eating probably contributes largely to the outstanding flavour release properties of these gels. It has been claimed that the improved flavour release provides an opportunity to use less flavour or less sugar in a dessert product gelled with gellan gum (Owen, 1989).

More comprehensive texture–flavour relationship testing suggests that a relationship exists between gel hardness and, to a lesser extent, brittleness and overall flavour (Table 10.4) (Clark, 1990). The latter increases as gel hardness decreases and gels become increasingly brittle. The ability of gellan gum to produce highly brittle gels is therefore of value in promoting good flavour properties. More importantly, its ability to produce gels with a high modulus value, which are perceived as very firm, allows gels to be produced at lower gum concentration with consequent reduction in gel hardness and greater flavour benefit.

10.4.5 Gum combinations

10.4.5.1 Gellan gum with thickeners. It is common practice to include thickeners in gel systems to reduce syneresis, improve freeze–thaw

Table 10.4 Relationship between flavour, hardness and brittleness. (From Clark, 1990.)

Hardness (kgf)	Brittleness (%)	Overall flavour
0.73	36.4	37.7
1.45	37.1	34.5
3.09	36.6	20.7
3.09	64.2	17.4
1.45	44.5	23.5

stability or, in some cases, to prevent unfavourable interactions between ingredients. For the same reasons it may be advantageous to include a thickener in gellan gum systems.

Thickeners such as xanthan gum, guar gum, locust bean gum or sodium carboxymethylcellulose do not appreciably change the texture of gellan gum gels. For example, at a constant total gum concentration, increasing the proportion of xanthan gum produces the changes that would be anticipated from the reduction in gellan gum concentration, namely reduction in gel strength with neglible change in the flexural parameters.

10.4.5.2 Gellan gum with other gelling agents. Blending gelling agents to modify gel texture is less commonly practised, and the mechanisms involved are little understood. While their gel structure is complex, and various possible networks have been proposed, the resulting gel texture may be predicted from a knowledge of the texture of the individual gelling components. Thus, blending gellan gum with other thermosetting gelling agents produces predictable effects.

Agar and κ-carrageenan also produce brittle gels. When used in combination with gellan gum, gel strength is reduced but the brittle texture of gellan gum gels is not changed. On the other hand, gelling agents which yield soft, elastic gels, such as xanthan gum–locust bean gum blends or the substituted form of gellan gum, produce a wide spectrum of gel textures when blended with gellan gum. These range from hard and brittle to soft and elastic, depending on the proportion of each gelling agent.

10.4.5.3 Gellan gum with starch. Starch, in its various forms, also has a predictable effect on the texture of gellan gum gels but, as it is the most widely used food hydrocolloid, it is perhaps more pertinent to consider how its properties are influenced by small additions of gellan gum. Starches impart a thick, paste-like consistency or in some cases a gel-like structure to foods, and other hydrocolloids are often used to modify this texture or reduce syneresis (Figure 10.16). In contrast to hydrocolloids

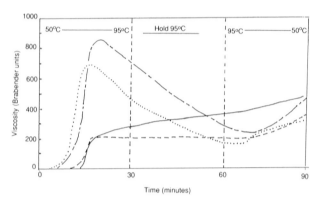

Figure 10.16 Brabender amylographs for starch (4.5%) + thickener (0.5%). (After Sanderson *et al.*, 1987.) —, starch; ···, starch + xanthan gum; ———, starch + carboxymethylcellulose; ————, starch + gellan gum.

such as xanthan gum or sodium carboxymethylcellulose, which cause a significant increase in peak viscosity during preparation of a starch paste, gellan gum does not markedly alter the viscosity behaviour. On cooling, the presence of gellan gum above 0.1% confers a firmer, shorter texture (Sanderson *et al.*, 1987b). In practical terms this allows the level of starch used in some puddings and pie fillings to be reduced significantly with consequent flavour release benefits.

10.4.5.4 Gellan gum with gelatin. Gelatin is the most widely used gelling hydrocolloid. In contrast to the firm, brittle, non-elastic gels produced by gellan gum, gelatin yields very elastic, non-brittle gels with very little perceived firmness. Combinations of gellan gum with gelatin can therefore produce a spectrum of textures depending on the relative proportion of the individual gelling agents (Wolf *et al.*, 1989). Unlike blends of gellan gum with xanthan and locust bean gums, which yield gels whose strength is intermediate between that of the individual gelling agents, combinations of gellan gum with gelatin can produce an increase in gel strength while still yielding anticipated flexural changes.

In common with all anionic polysaccharide–protein interactions, the effects achieved by blending gellan gum with gelatin depend up a number of factors, including pH, temperature, ionic strength, time, total and relative hydrocolloid concentrations and gelatin type. Manipulation of these factors can lead to gels with the range of textures described or can lead to precipitation of the two hydrocolloids. Under conditions in which gellan gum and gelatin precipitate, it is possible to induce coacervation and the formation of microcapsules (Chilvers and Morris, 1987).

Under conditions in which compatibility is achieved, the inclusion of a

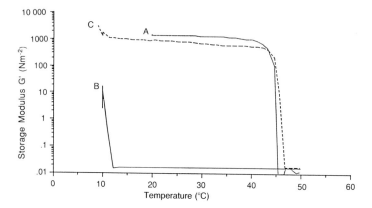

Figure 10.7 Setting behaviour of a gellan gum/gelatin blend (3:40) at pH 5: A, gellan gum; B, gelatin; C, gellan gum/gelatin blend.

small proportion of gellan gum produces an initial higher setting temperature for the gelatin gel, which is useful in applications where further processing is required such as in the preparation of multilayer desserts (Figure 10.17). Preliminary studies, in which the development of gel structure was monitored by the increase in shear modulus of the system with reducing temperature, clearly showed two distinct setting temperatures which coincide with the setting of the individual gellan gum and gelatin components. The mixed gel therefore probably consists of a gelatin gel entrapped within a gellan gum gel structure. Further work is required to substantiate these initial findings.

Limited model studies of combinations with other proteins have shown that at neutral pH gellan gum is compatible with milk proteins, soy, egg albumen, whey and sodium caseinate, but precipitation with all of these proteins occurs at about pH 4. Notwithstanding the results of model studies, it has been shown that it is possible to produce a number of directly acidified dairy products using gellan gum in combination with protective hydrocolloids such as guar gum or sodium carboxymethylcellulose, thus emphasising the need to study gellan gum–protein interactions under specific in-use conditions.

10.6 Availability and regulatory status

Approval for the use of gellan gum as a food additive is being sought worldwide. It is permitted in Japan, where it has been on sale since 1989. In the USA it is recognised as a food additive under the provisions of the Food and Drug Administration regulations (21CFR 172.665 issued in the

Table 10.5 Potential application areas for gellan gum.

Potential application area	Typical products	Typical use level (%)
Water-based jellies	Dessert jellies, aspics	0.15–0.2
Jams and jellies	Low-calorie spreads, imitation jams, bakery fillings	0.12–0.3
Confectionery	Pastille-type confectionery, marshmallows	0.8–1.0
Fabricated foods	Fabricated fruit, vegetables or meats	0.2–0.3
Icings	Bakery icings, frostings	0.05–0.12
Dairy products	Milk desserts	0.1–0.2
Pie fillings	Fruit pie fillings	0.25–0.35

Federal Register 28.9.90) for use as a stabiliser and thickener in icings, frostings, glazes and non-standardised jams and jellies.

The European Community's Scientific Committee for Food and the World Health Organization's Joint Expert Committee on Food Additives (JECFA) have independently given gellan gum a safety clearance, both granting an acceptable daily intake (ADI) of 'not specified'.

10.7 Potential applications

Gellan gum is still a comparatively new gelling agent and, even after it becomes widely permitted as a food additive, widespread use will inevitably take time. The properties described herein should nevertheless alert food technologists to its potential as a partial or total replacement of existing gelling agents and as an essential tool in the development of new food products.

The high efficiency and compatibility of gellan gum with other food ingredients, together with the ease with which it can fit into many existing processes, are important factors in its broad potential application in foods. These points are summarised in Table 10.5. The level of gellan gum used in these systems generally ranges from 0.02% in icings to 0.4% for the firm structure of a Japanese red bean jelly, or exceptionally up to about 1% in a high-solids (80%) confectionery jelly such as a fruit pastille. Gellan gum offers confectionery manufacturers the opportunity to make jelly sweets by partial or total replacement of other gelling agents while maintaining their existing process, but with improved acid stability during processing. Alternatively, a gellan gum-based product can be formulated to set in a very short time with the potential elimination of stoving. A sample recipe for a gellan gum confection is given in Table 10.6.

The low-solids fruit spread whose formulation is given in Table 10.7

Table 10.6 Sample recipe for a fruit pastille-type gel.

	Percentage of mix at 76% solids
Sucrose	40.4
Glucose syrup solids (42 DE)	32.2
Gellan gum	0.8
Sodium citrate	1.4
Citric acid	1.4
Colour and flavour	As required

DE, dextrose equivalent

Table 10.7 Sample recipe for a low-solids fruit spread.

Ingredient	Content (%)
Fruit	55.0
Sugar	26.3
Water	17.35
Citric acid	0.8
Gellan gum	0.5
Potassium sorbate	0.05

Table 10.8 Sample recipe for a dessert jelly.

Ingredient	Content (%)
Water	84.18
Sugar	15.01
Citric acid anhydrous	0.40
Trisodium citrate dihydrate	0.25
Gellan gum	0.16
Colour and flavour	As required

has been produced on existing pilot-scale equipment with trouble-free processing. The product has good appearance, spreadability and flavour release. Additionally, processing tolerance is enhanced, as temperature variations and overheating caused by extended pasteurisation or holding in heated tanks prior to filling have little impact on the final product.

Water-based, fruit-flavoured jellies are probably the simplest examples of gellan gum application. Such a product can be formulated as a powder dessert (Table 10.8) for domestic preparation or as a ready-to-eat jelly. While this simple jelly is texturally different from a gelatin jelly, it has certain advantages:

Table 10.9 Sample recipe for a milk jelly.

Ingredient	Content (%)
Milk	86.43
Sugar	13.10
Disodium hydrogen orthophosphate	0.28
Gellan gum	0.11
Salt	0.08
Colour and vanilla flavour	As required

- More rapid setting
- Unaffected by changes in storage temperature
- Enhanced processing tolerance
- Improved flavour release
- More refreshing taste.

These are particularly important for ready-to-eat products, for which there is a substantial market in Japan and the Far East.

While the applications described above are all based on make-up in water, gellan gum gels can be prepared just as easily in milk. Table 10.9 gives a formulation for a simple milk-based analogue of the afore-mentioned water jelly. Again the product can be sold either as a powder mix for instant preparation or as a ready-to-eat dessert. The texture of the milk dessert can conveniently be modified by increasing the gellan gum concentration or by the inclusion of 1–2% of a suitable starch.

These few examples illustrate a wide range of potential applications for gellan gum. Its good thermal and acid stability are major benefits in the production of all of these products, while its efficiency as a gelling agent allows its use at a fraction of the level required by other gelling agents.

10.8 Future developments

Research and development is continuing to extend the range of potential applications for gellan gum and, with an increasing consumer demand for healthy foods, much of this will focus on the formulation of low-calorie products.

Gel texture variations will continue to be obtained by appropriate blending with other gelling agents. However, as the substituted form of gellan becomes commercially available it will provide the opportunity for blending with the unsubstituted form to satisfy the textural requirements of food formulators.

Gellan gum will also form strong, brittle films when solutions are dried and these can be used to modify the tensile properties of other more flexible films. Plasticisers may be incorporated to obtain films with a wide range of properties. Just as combinations of proteins and lipids can form moisture barriers, the anionic gellan gum can be emulsified with water-insoluble fatty amines to produce films with moisture barrier properties. In this respect, tests with a range of fatty amines have shown that, as the hydrocarbon chain of the fatty amine increases, the water permeability decreases to a minimum at a chain length of 12 carbon atoms. For example, a film formed from an emulsion containing gellan gum (2 parts), lauryl amine (4 parts) and gelatin (2 parts) shows a water vapour transmission over 6 h of 8% compared with 84% through a film formed from an aqueous solution of gellan gum (2.5 parts) and gelatin (2.5 parts).

The anionic character of gellan gum can be changed by esterification. To this end a propylene glycol ester has been prepared which has been shown to have weak surface activity. In some model systems the propylene glycol gellan gum was found to be superior to propylene glycol alginate, and in others it was found to be a little inferior. While these properties will undoubtedly depend upon the degree of esterification and improved products can be prepared, it is unlikely that this avenue will be pursued in view of the high cost of establishing food use approval for any new polysaccharide derivative.

References

Chandrasekaran, R., Millane, R.P., Arnott, S. and Atkins, E.D.T. (1988a) The crystal structure of gellan. *Carbohydr. Res.*, **175**, 1–15.

Chandrasekaran, R., Puigjaner, L.C., Joyce, K.L. and Arnott, S. (1988b) Cation interactions in gellan: an X-ray study of the potassium salt. *Carbohydr. Res.*, **181**, 23–40.

Chilvers, G.R. and Morris, V.J. (1987) Coacervation of gelatin–gellan gum mixtures and their use in micro-encapsulation. *Carbohydr. Polym.*, **7**, 111–120.

Clark, R.C. (1990) Flavour and texture factors in model gel systems. In: *Food Technology International, Europe*, A. Turner, ed, Sterling Publications International, pp. 272–277.

Grasdalen, H. and Smidsrod, O. (1987) Gelation of gellan gum. *Carbohydr. Res.*, 371–393.

Gunning, A.P. and Morris, V.J. (1990) Light scattering studies of tetramethyl ammonium gellan. *Int. J. Biol. Macromol.*, **12**, 338–341.

Kang, K.S. and Veeder, G.T. (1982) Polysaccharide S-60 and bacterial fermentation process for its preparation. US Patent 4,326,053.

Kang, K.S. and Veeder, G.T. (1983) Fermentation process for preparation of polysaccharide S-60. US Patent 4,377,636.

Kang, K.S., Colgrove, G.T. and Veeder, G.T. (1980) Heteropolysaccharides produced by bacteria and derived products. European Patent 12,552.

Kang, K.S., Veeder, G.T., Mirrasoul, P.J., Kaneko, T. and Cottrell, I.W. (1981) Agar-like polysaccharide produced by a *Pseudomonas* species: production and basic properties. *Appl. Environm. Microbiol.*, **4(5)**, 1086–1091.

Kang, K.S., Colgrove, G.T. and Veeder, G.T. (1982) De-acetylated polysaccharide S-60. US Patent 4,326,052.

Kang, K.S., Veeder, G.T. and Colgrove, G.T. (1983) De-acetylated polysaccharide S-60. US Patent 4,385,125.

Kuo, M-S., Mort, A.J. and Dell, A. (1986) Identification and location of L-glycerate, an unusual acyl substituent in gellan gum. *Carbohydr. Res.*, **156**, 173–187.

Owen, G. (1989) Gellan gum-quick setting gelling systems. In: *Gums and Stabilisers for the Food Industry*, Vol. 5, G.O. Phillips, D.J. Wedlock and P.A. Williams, eds, IRL Press at Oxford University Press, Oxford, pp. 345–349.

Robinson, G., Manning, C.E., Morris, E.R. and Dea, I.C.M. (1987) Sidechain and mainchain interactions in bacterial polysaccharides. In: *Gums and Stabilisers for the Food Industry*, Vol. 4, G.O. Phillips, D.J. Wedlock and P.A. Williams, eds, IRL Press, Oxford, pp. 173–181.

Sanderson, G.R., Bell, V.L., Clark, R.C. and Ortega, D. (1987a) The texture of gellan gum gels. In: *Gums and Stabilisers for the Food Industry* Vol. 4, G.O. Phillips, D.J. Wedlock and P.A. Williams, eds, IRL Press, Oxford, pp. 219–229.

Sanderson, G.R., Bell, V.L., Burgum, D.R., Clark, R.C. and Ortega, D. (1987b) Gellan gum in combination with other hydrocolloids. In: *Gums and Stabilisers for the Food Industry*, Vol. 4 G.O. Phillips, D.J. Wedlock and P.A. Williams, eds, IRL Press, Oxford, pp. 301–308.

Wolf, C.L., LaVelle, W.M. and Clark, R.C. (1989) Gellan gum/gelatin blands, US Patent 4,876,105.

Sanderson, G.R. (1989) The functional properties and applications of microbial polysaccharides — a supplier's view. In: *Gums and Stabilisers for the Food Industry*, Vol. 5, G.O. Phillips, D.J. Wedlock and P.A. Williams, eds, IRL Press at Oxford University Press, Oxford, pp. 333–344.

Index